ASE Test Preparation

Automobile Certification Series

Automatic Transmission/Transaxle (A2)

5th Edition

DELMAR
CENGAGE Learning™

Australia • Brazil • Japan • Korea • Mexico • Singapore • Spain • United Kingdom • United States

DELMAR
CENGAGE Learning™

ASE Test Preparation: Automobile Certification Series, Automatic Transmission/Transaxle (A2), 5th Edition

Vice President, Technology and Trades Professional Business Unit: Gregory L. Clayton

Director, Professional Transportation Industry Training Solutions: Kristen L. Davis

Product Manager: Lori Bonesteel

Editorial Assistant: Danielle Filippone

Director of Marketing: Beth A. Lutz

Marketing Manager: Jennifer Barbic

Senior Production Director: Wendy Troeger

Senior Art Director: Benjamin Gleeksman

Content Project Management: PreMediaGlobal

Section Opener Image: Image Copyright Arcobalena, 2012. Used under license from Shutterstock.com

For product information and technology assistance, contact us at
Cengage Learning Customer & Sales Support, 1-800-354-9706

For permission to use material from this text or product, submit all requests online at **www.cengage.com/permissions**. Further permissions questions can be e-mailed to **permissionrequest@cengage.com**

ISBN-13: 978-1-111-12704-6

ISBN-10: 1-111-12704-2

Delmar Cengage Learning
5 Maxwell Drive
Clifton Park, NY 12065-2919
USA

Cengage Learning is a leading provider of customized learning solutions with office locations around the globe, including Singapore, the United Kingdom, Australia, Mexico, Brazil, and Japan. Locate your local office at: **international.cengage.com/region**

Cengage Learning products are represented in Canada by Nelson Education, Ltd.

For more information on transportation titles available from Delmar Cengage Learning, please visit our website at **www.trainingbay.cengage.com**

For more learning solutions, please visit our corporate website at **www.cengage.com**

Notice to the Reader

Publisher does not warrant or guarantee any of the products described herein or perform any independent analysis in connection with any of the product information contained herein. Publisher does not assume, and expressly disclaims, any obligation to obtain and include information other than that provided to it by the manufacturer. The reader is expressly warned to consider and adopt all safety precautions that might be indicated by the activities described herein and to avoid all potential hazards. By following the instructions contained herein, the reader willingly assumes all risks in connection with such instructions. The publisher makes no representations or warranties of any kind, including but not limited to, the warranties of fitness for particular purpose or merchantability, nor are any such representations implied with respect to the material set forth herein, and the publisher takes no responsibility with respect to such material. The publisher shall not be liable for any special, consequential, or exemplary damages resulting, in whole or part, from the readers' use of, or reliance upon, this material.

Printed in the United States of America
1 2 3 4 5 6 7 15 14 13 12 11

Table of Contents

Delmar, a part of Cengage Learning, is very pleased that you have chosen to use our ASE Test Preparation Guide to help prepare yourself for the Automatic Transmission/Transaxle (A2) ASE Certification Examination. This guide is designed to help prepare you for your actual exam by providing you with an overview and introduction of the testing process, introducing you to the task list for the Automatic Transmission/Transaxle (A2) certification exam, giving you an understanding of what knowledge and skills you are expected to have in order to successfully perform the duties associated with each task area, and providing you with several preparation exams designed to emulate the live exam content in hopes of assessing your overall exam readiness.

If you have a basic working knowledge of the discipline you are testing for, you will find this book to be an excellent guide, helping you to understand the "must know" items needed to successfully pass the exam. This book is not a textbook. Its objective is to prepare the technician who has the requisite experience and schooling to take on the challenge of the ASE certification process. This guide cannot replace the hands-on experience and theoretical knowledge required by ASE to master the vehicle repair technology associated with this exam. If you are unable to understand more than a few of the preparation questions and their corresponding explanations in this book, it could be that you require either more shop-floor experience or further study.

This book begins by providing an overview and introduction of the testing process. This section will outline what we recommend you do to prepare, what to expect on the actual test day, and overall methodologies for your success. This section is followed by a detailed overview of the ASE task list to include explanations of the knowledge and skills you must possess to successfully answer questions related to each particular task. After the task list, we provide six sample preparation exams for your use as a means of evaluating areas of understanding, as well as areas requiring improvement in order to successfully pass the ASE exam. Delmar is the first and only test preparation organization to provide so many unique preparation exams. We enhanced our guides to include this much support as a means of providing you with the best preparation product available. Section 6 of this guide includes the answer key preparation exam, along with the answer explanations for each question. Each answer explanation also contains a reference back to the related task or tasks that it assesses. This will provide you with a quick and easy method for referring back to the task list whenever needed. The last section of this book contains blank answer sheet forms you can use as you attempt each preparation exam, along with a glossary of terms.

OUR COMMITMENT TO EXCELLENCE

Thank you for choosing Delmar, Cengage Learning for your ASE test preparation needs. All of the writers, editors, and Delmar staff have worked very hard to make this test preparation guide second to none. We feel confident that you will find this guide easy to use and extremely beneficial as you prepare for your actual ASE exam.

Delmar, Cengage Learning has sought out the best subject-matter experts in the country to help with the development of *ASE Test Preparation: Automobile Certification Series, Automatic Transmission/Transaxle (A2), 5th Edition*. Preparation questions are authored and then reviewed by a group of certified, subject-matter experts to ensure the highest level of quality and validity to our product.

If you have any questions concerning this guide or any guide in this series, please visit us on the web at **http://www.trainingbay.cengage.com**.

For web-based online test preparation for ASE certifications, please visit us on the web at **http://www.techniciantestprep.com/** to learn more.

ABOUT THE AUTHOR

Jerry Clemons has been around cars, trucks, equipment, and machinery throughout his whole life. Being raised on a large farm in central Kentucky provided him with an opportunity to complete mechanical repair procedures from an early age. Jerry earned an associate in applied science degree in Automotive Technology from Southern Illinois University and a bachelor of science degree in Vocational, Industrial, and Technical Education from Western Kentucky University. Jerry has also completed a master of science degree in Safety, Security, and Emergency Management from Eastern Kentucky University. Jerry has been employed at Elizabethtown Community and Technical College since 1999 and is currently an associate professor for the Automotive and Diesel Technology Programs. Jerry holds the following ASE certifications: Master Medium/Heavy Truck Technician, Master Automotive Technician, Advanced Engine Performance (L1), Truck Equipment Electrical Installation (E2), and Automotive Service Consultant (C1). Jerry is a member of the Mobile Air Conditioning Society (MACS), as well as a member of the North American Council of Automotive Teachers (NACAT). Jerry has been involved in developing transportation material for Cengage Learning for seven years.

ABOUT THE SERIES ADVISOR

Mike Swaim has been an Automotive Technology Instructor at North Idaho College, Coeur d'Alene, Idaho, since 1978. He is an Automotive Service Excellence (ASE) Certified Master Technician since 1974 and holds a Lifetime Certification from the Mobile Air Conditioning Society. He served as Series Advisor to all nine of the 2011 Automobile/Light Truck Certification Tests (A Series) of Cengage, Delmar Learning ASE Test Preparation titles, and is the author of *ASE Test Preparation: Automobile Certification Series, Undercar Specialist Designation (X1), 5th Edition*.

ASE began as the National Institute for Automotive Service Excellence (NIASE). It was founded as a nonprofit, independent entity in 1972 by a group of industry leaders with the single goal of providing a means for consumers to distinguish between incompetent and competent technicians. It accomplishes this goal through the testing and certification of repair and service professionals. Though it is still known as the National Institute for Automotive Service Excellence, it is now called "ASE" for short.

Today, ASE offers more than 40 certification exams in automotive, medium/heavy-duty truck, collision repair and refinish, school bus, transit bus, parts specialist, automobile service consultant, and other industry-related areas. At this time there are more than 385,000 professionals nationwide with current ASE certifications. These professionals are employed by new car and truck dealerships, independent repair facilities, fleets, service stations, franchised service facilities, and more.

ASE's certification exams are industry driven and cover practically every on-highway vehicle service segment. The exams are designed to stress the knowledge of job-related skills. Certification consists of passing at least one exam and documenting two years of relevant work experience. To maintain certification, those with ASE credentials must be re-tested every five years.

While ASE certifications are a targeted means of acknowledging the skills and abilities of an individual technician, ASE also has a program designed to provide recognition for highly qualified repair, support, and parts businesses. The Blue Seal of Excellence Recognition Program allows businesses to showcase their technicians and their commitment to excellence. One of the requirements of becoming Blue Seal recognized is that the facility must have a minimum of 75 percent of its technicians ASE certified. Additional criteria apply, and program details can be found on the ASE website.

ASE recognized that educational programs serving the service and repair industry also needed a way to be recognized as having the faculty, facilities, and equipment to provide a quality education to students wanting to become service professionals. Through the combined efforts of ASE, industry, and education leaders, the nonprofit National Automotive Technicians Education Foundation (NATEF) was created in 1983 to evaluate and recognize academic programs. Today, more than 2,000 educational programs are NATEF certified.

For additional information about ASE, NATEF, or any of their programs, the following contact information can be used:

National Institute for Automotive Service Excellence (ASE)

101 Blue Seal Drive S.E.

Suite 101

Leesburg, VA 20175

Telephone: 703-669-6600

Fax: 703-669-6123

Website: **www.ase.com**

Overview and Introduction

Participating in the National Institute for Automotive Service Excellence (ASE) voluntary certification program provides you with the opportunity to demonstrate you are a qualified and skilled professional technician that has the "know-how" required to successfully work on today's modern vehicles.

EXAM ADMINISTRATION

Through 2011, there are two methods available to you when taking an ASE certification exam:

- Paper and pencil
- Computer-based testing (CBT)

> *Note:* Beginning 2012, ASE will no longer offer paper and pencil certification exams. It will offer and support CBT testing exclusively.

Paper and Pencil Exams

ASE paper and pencil exams are administered twice annually, once in the spring and once again in the fall. The paper and pencil exams are administered at over 750 exam sites in local communities across the nation.

Each test participant is given a booklet containing questions with charts and diagrams where required. All instructions are printed on the exam materials and should be followed carefully. You can mark in this exam booklet but no information entered in the booklet is scored. You will record your answers using a separate answer sheet. You will need to mark your answers using only a number 2 pencil. Upon completion of your exam, the answer sheets are electronically scanned and the answers are tabulated.

> *Note:* Paper and pencil exams will no longer be offered by ASE after 2011. ASE will be converting to a completely exclusive CBT testing methodology at that time.

CBT Exams

ASE also provides computer-based testing (CBT) exams that are administered twice annually, once in the winter and once again in the summer. The CBT exams are administered at test centers across the nation. The exam content is the same for both the paper and pencil and CBT testing methods.

If you are considering the CBT exams, it is recommended that you go to the ASE website at *http://www.ase.com* and review the conditions and requirements for this type of exam. There is also an exam demonstration page that allows you to personally experience how this type of exam operates before you register.

Effective 2012, ASE will only offer CBT testing. At that time, CBT exams will be available four times annually for two-month windows, with a month of no testing in between each testing window.

- January/February – Winter CBT testing window
- April/May – Spring CBT testing window
- July/August – Summer CBT testing window
- October/November – Fall CBT testing window

Please note, testing windows and timing may change. It is recommended you go to the ASE website at *http://www.ase.com* and review the latest testing schedules.

UNDERSTANDING TEST QUESTION BASICS

ASE exam questions are written by service industry experts. Each question on an exam is created during an ASE-hosted "item-writing" workshop. During these workshops, expert service representatives from manufacturers (domestic and import), aftermarket parts and equipment manufacturers, working technicians, and technical educators gather to share ideas and convert them into actual exam questions. Each exam question written by these experts must then survive review by all members of the group. The questions are designed to address the practical application of repair and diagnosis knowledge and skills practiced by technicians in their day-to-day work.

After the item-writing workshop, all questions are pretested and quality checked on a national sample of technicians. Those questions that meet ASE standards of quality and accuracy are included in the scored sections of the exams; the "rejects" are sent back to the drawing board or discarded altogether.

Depending on the topic of the certification exam, you will be asked between 40 and 80 multiple-choice questions. You can determine the approximate number of questions you can expect to be asked during the Automatic Transmission/Transaxle (A2) certification exam by reviewing the task list in Section 4 of this book. The five-year recertification exam will cover this same content; however, the number of questions for each content area of the recertification exam will be reduced by approximately one-half.

> *Note:* Exams may contain questions that are included for statistical research purposes only. Your answers to these questions will not affect your score, but since you do not know which ones they are, you should answer all questions in the exam.

Using multiple criteria, including cross sections by age, race, and other background information, ASE is able to guarantee that exam questions do not include bias for or against any particular group. A question that shows bias toward any particular group is discarded.

TEST-TAKING STRATEGIES

Before beginning your exam, quickly look over the exam to determine the total number of questions that you will need to answer. This knowledge will help you gauge your time throughout the exam to ensure you have enough available to answer all of the questions presented. Read through each question completely before marking your answer. Answer the questions in the order they appear on the exam. Leave the questions blank that you are not sure of and move on to the next question. You can return to those unanswered questions after you have finished the others. These questions may actually be easier to answer at a later time once your mind has had additional time to consider them on a subconscious level. In addition, you might find information in other questions that will help you recall the answers to some of them.

Multiple-choice exams are sometimes challenging because there are often several choices that may seem possible, or partially correct, and therefore it may be difficult to decide on the most appropriate answer choice. The best strategy, in this case, is to first determine the correct answer before looking at the answer options. If you see the answer you decided on, you should still be careful to examine the other answer options to make sure that none seems more correct than yours. If you do not know or are not sure of the answer, read each option very carefully and try to eliminate those options that you know are incorrect. That way, you can often arrive at the correct choice through a process of elimination.

If you have gone through the entire exam and you still do not know the answer to some of the questions, *then guess*. Yes, guess. You then have at least a 25 percent chance of being correct. While your score is based on the number of questions answered correctly, any question left blank or unanswered is automatically scored as incorrect.

There is a lot of "folk" wisdom on the subject of test taking that you may hear about as you prepare for your ASE exam. For example, there are those who would advise you to avoid response options that use certain words such as *all, none, always, never, must,* and *only,* to name a few. This, they claim, is because nothing in life is exclusive. They would advise you to choose response options that use words that allow for some exception, such as *sometimes, frequently, rarely, often, usually, seldom,* and *normally.* They would also advise you to avoid the first and last option (A or D) because exam writers, they feel, are more comfortable if they put the correct answer in the middle (B or C) of the choices. Another recommendation often offered is to select the option that is either shorter or longer than the other three choices because it is more likely to be correct. Some would advise you to never change an answer since your first intuition is usually correct. Another area of folk wisdom focuses specifically on any repetitive patterns created by your question responses (for example, A, B, C, A, B, C, A, B, C).

Many individuals may say that there are actual grains of truth in this folk wisdom, and whereas with some exams this may prove true, it is not relevant in regard to the ASE certification exams. ASE validates all exam questions and test forms through a national sample of technicians, and only those questions and test forms that meet ASE standards of quality and accuracy are included in the scored sections of the exams. Any biased questions or patterns are discarded altogether. Therefore, it is highly unlikely you will actually experience any of this folk wisdom on an actual ASE exam.

PREPARING FOR THE EXAM

Delmar, Cengage Learning wants to make sure we are providing you with the most thorough preparation guide possible. To demonstrate this, hundreds of preparation questions are included in this guide. These questions are designed to provide as many opportunities as possible to prepare you to successfully attempt and pass your ASE exam. The preparation approach recommended and outlined in this book is designed to help you build confidence in demonstrating what task area content you already know well while also outlining what areas you should review in more detail prior to the actual exam.

We recommend that your first step in the preparation process should be to thoroughly review Section 3 of this book. This section contains a description and explanation of the type of questions you'll find on an ASE exam.

Once you understand how the questions will be presented, we then recommend that you thoroughly review Section 4 of this book. This section contains information that will help you establish an understanding of what the exam will be evaluating, and specifically, how many questions to expect in each specific task area.

As your third preparatory step, we recommend you attempt your first preparation exam, located in Section 5 of this book. Answer one question at a time. After you answer each question, review the answer and question explanation information located in Section 6. This section will provide you with instant response feedback allowing you to gauge your progress, one question at a time, throughout this first preparation exam attempt. If after reading the question explanation you do not feel you understand the reasoning for the correct answer, go back and review the task list overview (Section 4) for the task that is related to that question. Included with each question explanation, there is a clear identifier of the task area that is being assessed (for example, Task A.1). If at that point you still don't feel you have a solid understanding of the material, identify a good source of information on the topic, such as an educational course, textbook, or other related source of topical learning, and do some additional studying.

After you have completed your first preparation exam and have reviewed your answers, you are ready to attempt your next preparation exam. A total of six practice exams are available in Section 5 of this book. For your second preparation exam, we recommend that you answer the questions as if you were taking the actual exam. Do not use any reference material or allow any interruptions in order to get a feel for how you will do on the actual exam. Once you have answered all of the questions, grade your results using the answer key in Section 6. For every question that you gave an incorrect answer to study the explanations to the answers and/or the overview of the related task areas. Try to determine the root cause for your missing the question. The easiest thing to correct is learning the correct technical content. The hardest things to correct are behaviors that lead you to an incorrect conclusion. If you knew the information but still got the question incorrect there is likely a test-taking behavior that needs to be corrected. An example of this would be reading too quickly and skipping over words that affect your reasoning. If you can identify what you did that caused you to answer the question incorrectly, you can eliminate that cause and improve your score.

Here are some basic guidelines to follow while preparing for the exam:

- Focus your studies on those areas you are weak in.
- Be honest with yourself when determining if you understand something.
- Study often but for short periods of time.
- Remove yourself from all distractions when studying.
- Keep in mind the goal of studying is not just to pass the exam; the real goal is to learn.
- Prepare physically by getting a good night's rest before the exam and eat meals that provide energy but do not cause discomfort.
- Arrive early to the exam site to avoid long waits as test candidates check in.
- Use all of the time available for your exams. If you finish early, spend the remaining time reviewing your answers.
- Do not leave any questions unanswered. If absolutely necessary, guess. All unanswered questions are automatically scored as incorrect.

Here are some items you will need to bring to the exam site:

- A valid, government- or school-issued photo ID
- Your test center admissions ticket
- Three or four sharpened #2 pencils and an eraser
- A watch (not all test sites have clocks)

Note: Books, calculators, and other reference material are not allowed in the exam room. The exceptions to this list are English-Foreign dictionaries, or glossaries. All items will be inspected before and after testing.

WHAT TO EXPECT DURING THE EXAM

Paper and Pencil Exams

When taking a paper and pencil exam, you will be placing your answers on a sheet that requires you to blacken (bubble) in your answer choice.

Be careful that only your answers are visible on the answer sheet. Stray pencil marks or incomplete erasures may be picked up as an answer by the electronic reader and result in a question being scored incorrectly.

Studies have shown that one of the biggest challenges an adult faces when taking a test that uses a bubble-style answer sheet is in placing answers in the correct location. To avoid having problems in this area, be extra mindful of how and where you mark your answers. For example, when answering question 21, blacken the correct, corresponding bubble on the answer sheet for question 21. Pay special attention to this process when you decide to skip a question to come back to later. In this situation, many people forget to also leave the corresponding line on the bubble answer sheet blank as well. They inadvertently place their answer for the next question on the answer bubble sheet line that should have been left as a blank placeholder for the unanswered, skipped question. Providing a correct question response on the incorrect bubble answer sheet line will likely result in that question being marked wrong. Remember, the answer sheet for the paper and pencil exam is machine scored and can only "read" what you have blackened or bubbled in.

If you finish answering all of the questions on an exam and have remaining time, go back and review the answers for those questions that you were not sure of. You can often catch careless errors by using the remaining time to review your answers. Carefully check your answer sheet for blank answers or missing information.

At practically every exam, some technicians will finish ahead of time and turn their papers in long before the final call. Since some technicians may be doing a recertification test and others may be taking fewer exams than you, do not let this distract or intimidate you.

It is not wise to use less than the total amount of time that you are allotted for an exam. If there are any doubts, take the time for review. Any product can usually be made better with some additional effort. An exam is no exception. It is not necessary to turn in your exam paper until you are told to do so.

CBT Exams

When taking a CBT exam, as soon as you are seated in the testing center, you will be given a brief tutorial to acquaint you with the computer-delivered test prior to taking your certification exam(s). Unlike paper and pencil testing, when attempting a CBT exam you will not have to worry about stray pencil marks or ensuring that your answers are marked on the correct and corresponding answer bubble sheet line. The CBT exams allow you to only select one answer per question. You can also change your answers as many times as you like. When you select a second answer choice, the CBT will automatically unselect your first answer choice. If you want to skip a question to return to later, you can utilize the "flag" feature, which allows you to quickly identify and review questions whenever you are ready. Prior to completing your exam, you will also be provided with an opportunity to review your answers and address any unanswered questions.

TESTING TIME

Paper and Pencil Exams

Each ASE paper and pencil exam session is four hours. You may register for and attempt anywhere from one to a maximum of four exams during any one exam session. It is recommended, however, that you do not attempt to register for any combination of exams that would result in you having to answer any more than 225 questions during any single exam session. As a worst-case scenario, this will allow you slightly more than one minute to answer each question.

CBT Exams

Unlike the ASE paper and pencil exams, each individual ASE CBT exam has a fixed time limit. Individual exam times will vary based upon exam area, and will range anywhere from a half hour to two hours. You will also be given an additional 30 minutes beyond what is allotted to complete your exams to ensure you have adequate time to perform all necessary check-in procedures, complete a brief CBT tutorial, and potentially complete a post-test survey.

Similar to the paper and pencil exams, you can register for and attempt multiple CBT exams during one testing appointment. The maximum time allotment for a CBT appointment is four and a half hours. If you happen to register for so many exams that you will require more time than this, your exams will be scheduled into multiple appointments. This could mean that you have testing on both the morning and the afternoon of the same day, or they could be scheduled on different days depending on your personal preference and the test center's schedule.

It is important to understand that if you arrive late for your CBT test appointment, you will not be able to make up any missed time. You will only have the scheduled amount of time remaining in your appointment to complete your exam(s).

Also, while most people finish their CBT exams within the time allowed, others might feel rushed or not be able to finish the test due to the implied stress of a specific, individual time limit allotment. Before you register for the CBT exams, you should review the number of exam questions that will be asked along with the amount of time allotted for that exam to determine whether you feel comfortable with the designated time limitation.

Summary

Regardless of whether you are taking a paper and pencil or CBT exam, as an overall time management recommendation you should monitor your progress and set a time limit you will follow in regard to how much time you will spend on each exam question. This should be based on the total number of questions you are attempting.

Also, it is very important to note that if for any reason you wish to leave the testing room during an exam, you must first ask permission. If you happen to finish your exam(s) early and wish to leave the testing site before your designated session appointment is completed, you are permitted to do so only during specified dismissal periods.

UNDERSTANDING HOW YOUR EXAM IS SCORED

You can gain a better perspective about the ASE certification exams if you understand how they are scored. ASE exams are scored by an independent organization having no vested interest in ASE or in the automotive industry.

Each question carries the same weight as any other question. For example, if there are 50 questions, each is worth 2 percent of the total score.

Your exam results can tell you:

- ■ Where your knowledge equals or exceeds that needed for competent performance, or
- ■ Where you might need more preparation.

Your ASE exam score report is divided into content "task" areas and will show the number of questions in each content area and how many of your answers were correct. These numbers provide information about your performance in each area of the exam. However, because there may be a different number of questions in each content area of the exam, a high percentage of correct answers in an area with few questions may not offset a low percentage in an area with many questions.

It should be noted that one does not "fail" an ASE exam. The technician who does not pass is simply told "More Preparation Needed." Though large differences in percentages may indicate problem areas, it is important to consider how many questions were asked in each area. Since each exam evaluates all phases of the work involved in a service specialty, you should be prepared in each area. A low score in one area could keep you from passing an entire exam.

There is no such thing as average. You cannot determine your overall exam score by adding the percentages given for each task area and dividing by the number of areas. It doesn't work that way because there generally are not the same number of questions in each task area. A task area with 20 questions, for example, counts more toward your total score than a task area with 10 questions.

Your exam report should give you a good picture of your results and a better understanding of your strengths and areas needing improvement for each task area.

If you fail to pass the exam, you may take it again at any time it is scheduled to be administered. You are the only one who will receive your exam score. Exam scores will not be given over the telephone by ASE nor will they be released to anyone without your written permission.

Understanding not only what content areas will be assessed during your exam, but how you can expect exam questions to be presented, will enable you to gain the confidence you need to successfully pass an ASE certification exam. The following examples will help you recognize the types of question styles used in ASE exams and assist you in avoiding common errors when answering them.

Both the paper and pencil and Computer Based Test (CBT) exam questions are identical in content and difficulty. Most initial certification tests are made up of between 40 and 80 multiple-choice questions. The five-year recertification exams will cover the same content as the initial exam; however, the actual number of questions for each content area will be reduced by approximately one-half. Refer to Section 4 of this book for specific details regarding the number of questions to expect to receive during the initial Automatic Transmission/Transaxle (A2) certification exam.

Multiple-choice questions are an efficient way to test knowledge. To correctly answer them, you must consider each answer choice as a possibility, and then choose the answer that *best* addresses the question. To do this, read each word of the question carefully. Do not assume you know what the question is asking until you finish reading the entire question.

About 10 percent of the questions on an actual ASE exam will reference an illustration. These drawings contain the information needed to correctly answer the question. The illustration should be studied carefully before attempting to answer the question. When the illustration shows a system in detail, look over the system and try to figure out how the system works before you look at the question and the possible answers. This approach will ensure you do not answer the question based upon false assumptions or partial data, but instead have reviewed the entire scenario being presented.

MULTIPLE-CHOICE QUESTIONS

The most common type of question used on an ASE exam is direct completion, which is more commonly referred to as a multiple-choice-style question. This type of question contains three "distracters" (incorrect answers) and one "key" (correct answer). When the questions are written, effort is made to make the distracters plausible to draw an inexperienced technician to inadvertently select one of them. This type of question gives a clear indication of the technician's knowledge.

Examples of this type of question would appear as follows:

1. Which of the following conditions could cause the transmission to slip when the vehicle makes sharp turns?

   A. Brake switch stuck open
   B. Engine ignition system misfire
   C. Low transmission fluid
   D. Restricted transmission cooler

Answer A is incorrect. A stuck open brake switch could cause the stop lights to be inoperative. A stuck open brake switch could cause the cruise control and/or the torque converter clutch to be inoperative.

Answer B is incorrect. An ignition system misfire could cause the vehicle to cut out under a load.

Answer C is correct. Low transmission fluid would cause the transmission to slip when the vehicle makes sharp turns due to the fluid moving away from the pick-up tube of the filter.

Answer D is incorrect. A restricted transmission cooler would cause the transmission fluid to overheat, which could cause transmission component failure from high heat levels.

2. Which of the following methods would most likely be used when testing the contact pattern on the final drive gears in a transaxle?

    A.   Gear compound

    B.   Slide caliper

    C.   Outside micrometer

    D.   Feeler gauge

Answer A is correct. Gear compound is often used when testing the contact pattern of the final drive gears in a transaxle. This substance provides good feedback about where the teeth are touching each other.

Answer B is incorrect. A slide caliper is used to make various transmission measurements such as outside diameter and inside diameter, as well as depth.

Answer C is incorrect. Outside micrometers are used to measure the outside diameter of components in the transmission.

Answer D is incorrect. A feeler gauge is used to check clutch pack clearance.

TECHNICIAN A, TECHNICIAN B QUESTIONS

The question style that is most popularly associated with an ASE exam is the "Technician A says … Technician B says … Who is correct?" type of question. In this type of question, you must identify the correct statement or statements. To answer this type of question correctly, you must carefully read each technician's statement and judge it on its own merit to determine if the statement is true.

Sometimes this type of question begins with a statement about some analysis or repair procedure. This is often referred to as the stem of the question and provides the setup or background information required to understand the conditions on which the question is based. This is followed by two statements about the cause of the concern, proper inspection, identification, or repair choices. You are asked whether the first statement, the second statement, both statements, or neither statement is correct. Analyzing this type of question is a little easier than the other types because there are only two ideas to consider, although there are still four choices for an answer.

Technician A, Technician B questions are really double true-or-false questions. The best way to analyze this type of question is to consider each technician's statement separately. Ask yourself, is A true or false? Is B true or false? Once you have completed this individual evaluation of each answer choice, you will have successfully determined the correct answer choice. An important point to remember is that an ASE Technician A, Technician B question will never have Technician A and B directly disagreeing with each other. That is why you must evaluate each statement independently.

An example of this type of question would appear as follows:

1. Technician A says that the transmission computer controls the torque converter lockup clutch. Technician B says that an inoperative torque converter lockup clutch can cause a decrease in fuel economy. Who is correct?

    A.   A only
    B.   B only
    C.   Both A and B
    D.   Neither A nor B

Answer A is incorrect. Technician B is also correct.

Answer B is incorrect. Technician A is also correct.

Answer C is correct. Both Technicians are correct. The transmission computer or the power train control module controls the operation of the torque converter clutch. Since the torque converter lockup clutch reduces the engine RPM when it engages, fuel economy would be reduced if the converter clutch was inoperative.

Answer D is incorrect. Both Technicians are correct.

EXCEPT QUESTIONS

Another question form used on the ASE exams contains answer choices that are all correct except for one. To help easily identify this type of question, whenever they are presented in an exam, the word "EXCEPT" will always be displayed in capital letters. With this type of question, the one incorrect answer choice will actually be counted as the correct answer for that question. Be careful to read these question types slowly and thoroughly, otherwise you may overlook what the question is actually asking and answer the question by selecting the first correct statement.

An example of this type of question would appear as follows:

1. All of the following conditions have to be met in order for the torque converter clutch to engage EXCEPT:

    A.   Light to moderate throttle is applied.
    B.   Vehicle is above the minimum converter clutch set speed.
    C.   Transmission is in second gear.
    D.   Engine temperature minimum is met.

Answer A is incorrect. The throttle application should be light to moderate in order for the torque converter clutch to engage.

Answer B is incorrect. The vehicle must be traveling above the minimum converter clutch set speed.

Answer C is correct. The torque converter will not typically be engaged in second gear. The torque converter clutch is usually engaged in the top two gears.

Answer D is incorrect. The engine temperature should be at least at the minimum level in order for the torque converter clutch to engage.

LEAST LIKELY QUESTIONS

For this type of question style, look for the answer choice that would be the LEAST LIKELY cause of the described situation. To help easily identify this type of question, whenever they are presented in an exam, the words "LEAST LIKELY" will always be displayed in capital letters. Read the entire question carefully before choosing your answer.

An example of this type of question would appear as follows:

1. Which of the following actions would LEAST LIKELY be performed by the technician while running a transmission fluid cooler flow test?

   A. Remove the cooler supply line.
   B. Remove the cooler return line.
   C. Start the engine and run at 1,000 RPM.
   D. Measure the fluid flow for 20 seconds.

Answer A is correct. The cooler supply line should not be removed while performing a cooler flow test.

Answer B is incorrect. The cooler return line should be removed while performing a cooler flow test.

Answer C is incorrect. The engine should be run at 1,000 RPM while performing a cooler flow test.

Answer D is incorrect. The cooler flow test is typically run for 20 seconds. One quart of fluid should be pumped out into the catch container in 20 seconds.

SUMMARY

The question styles outlined in this section are the only ones you will encounter on any ASE certification exam. ASE does not use any other types of question styles, such as fill-in-the-blank, true/false, word-matching, or essay. ASE also will not require you to draw diagrams or sketches to support any of your answer selections. If a formula or chart is required to answer a question, it will be provided for you.

Task List Overview

INTRODUCTION

This section of the book outlines the content areas or *task list* for this specific certification exam, along with a written overview of the content covered in the exam.

The task list describes the actual knowledge and skills necessary for a technician to successfully perform the work associated with each skill area. This task list is the fundamental guideline you should use to understand what areas you can expect to be tested on, and how each individual area is weighted to include the approximate number of exam questions you can expect to be given for that area during the ASE certification exam. It is important to note that the number of exam questions for a particular area is truly to be used as a guideline only. ASE advises that the questions on the exam may not equal the number specifically listed on the task list. The task lists are specifically designed to tell you what ASE expects you to know how to do and to help you be ready to be tested.

Similar to the role this task list will play in regard to the actual ASE exam, Delmar, Cengage Learning developed the six preparation exams, located in Section 5 of this book, using this task list as a guide. It is important to note that although both ASE and Delmar, Cengage Learning use the same task list as a guideline for creating our test questions, none of the test questions you will use in our practice exams will be found in the actual, live ASE exams. This is a true statement for any test preparatory material you use. Real exam questions are *only* visible during the actual ASE exams.

Task List at a Glance

The Automatic Transmission/Transaxle (A2) task list focuses on six core areas, and you can expect to be asked approximately 50 questions on your certification exam, broken out as outlined:

- A. General Transmission/Transaxle Diagnosis (25 questions)
    1. Mechanical/Hydraulic Systems (11)
    2. Electronic Systems (14)
- B. In-Vehicle Transmission/Transaxle Maintenance and Repair (12 questions)
- C. Off-Vehicle Transmission/Transaxle Repair (13 questions)
    1. Removal and Installation (4)
    2. Disassembly and Assembly (5)
    3. Friction and Reaction Units (4)

Based upon this information, the graph shown here is a general guideline demonstrating which areas will have the most focus on the actual certification exam. This data may help you prioritize your time when preparing for the exam.

Note: The actual number of questions you will be given on the ASE certification exam may vary slightly from the information provided in the task list, as exam contain questions that are included for statistical research purposes only. Do not forget that your answers to these research questions will not affect your score.

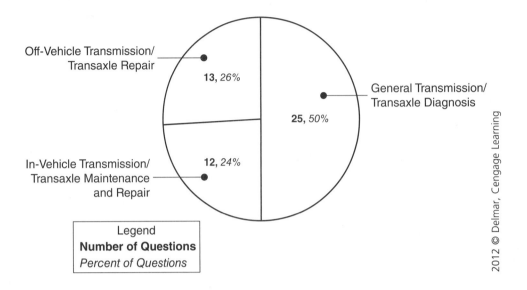

AUTOMATIC TRANSMISSION/TRANSAXLE (A2) TASK LIST

A. General Transmission/Transaxle Diagnosis (25 Questions)

1. Mechanical/Hydraulic Systems (11 Questions)

1. Road test the vehicle to verify mechanical/hydraulic system problems based on driver's concern; determine necessary action.

The most difficult part of verifying a customer's concern can be getting a good description of the problem. This task is usually left to the service consultant. A good consultant is invaluable to a technician. When armed with accurate problem occurrence conditions, such as vehicle hot or cold, type and speed of driving during problem occurrence, frequency at which problem occurs, and, if necessary, a road test with the customer, the technician can road test the vehicle to verify the problem and gain an understanding of it. The technician is trying to determine the root source of the problem. He is trying to decide if he is experiencing a problem related to the engine or the transmission. If the problem is transmission related, he uses his experience to pinpoint whether the problem is internally or externally rooted. Many transmissions can be tested using pressure gauges or scanners during the test drive to help the technician work toward identifying the root problem. Problem correction verification always requires test driving under occurrence conditions to insure proper repair.

2. Diagnose noise, vibration, harshness, and shift quality problems; determine necessary action.

In addition to the transmission and engine, the entire driveline should be checked before assuming a noise, vibration, or harshness problem is transmission related. Many engine speed-related vibrations are caused by an unbalanced torque converter assembly, a poorly mounted torque converter, or a faulty output shaft. The key to determining the cause of the vibration is to pay particular attention to the vibration in relationship to engine speed. If the vibration changes as vehicle speed changes, the causes could be the output shaft or the driveline connected to it.

Noise problems are also best diagnosed by paying a great deal of attention to the speed and condition at which the noise occurs. The conditions that merit the most attention are the operating gear and the load on the driveline. If the noise is related to engine speed and is present in all gears, including park and neutral, the most probable source of the noise is the oil pump. The noise could also be torque converter components, input shaft and attached components, and drive chain/sprockets, if equipped.

One method of pinpointing the source of the noise is varying the pump pressure at idle with transmission in park or neutral. Moving the throttle valve (TV) pressure control from maximum to minimum will vary pump output pressure, thereby changing the pitch of the noise. This can be further verified by watching the vibrations of an attached pressure gauge needle. If it has no effect, then the pump has been ruled out. Noises that occur only when a particular gear is operated must be related to those components responsible for providing that gear, such as a band or clutch. If the noise is related to vehicle speed, the most probable causes are the output shaft and final drive assembly. Often, the exact cause of the noise and vibration can only be identified through careful inspection of a disassembled transmission.

Shift quality issues can range from shift timing to condition problems. If a customer's concern is shifting too early or late, or too hard or too soft, does it happen in a specific gear or all gears? In nonelectronic shift applications, do the TV, kick-down, and/or vacuum modulator work correctly? Does the transmission or power train control module (PCM) have a strategy that causes hard or soft, or late or early, shifts under certain conditions or in different situations? Once you have verified the condition you must answer these questions to help guide you toward the root of the problem.

3. Diagnose fluid usage, type, level, and condition problems; determine necessary action.

The normal color of automatic transmission fluid (ATF) is red. If the fluid has a dark brownish or blackish color and/or burned odor, the fluid has been overheated. A milky fluid indicates that engine coolant has been leaking into the transmission's cooler in the radiator. If the fluid has a pink color, then the transmission has ingested water and all of the fluid will have to be exchanged with new fluid. After checking the ATF fluid level and color, wipe the dipstick on absorbent white paper and look at the stain left by the fluid. Dark particles are normally band and clutch material, and white, silvery, metal particles are normally caused by the wearing of the transmission's metal parts. If the dipstick cannot be wiped clean, it is probably covered with varnish, which results from fluid oxidation. Varnish or other heavy deposits indicate the need to change the transmission fluid and filter. Low fluid levels can cause a variety of problems. Air can be drawn into the oil pump's inlet circuit and mixed with the fluid. This will result in aerated fluid. This leads to slow pressure buildup and low pressures, which will cause slippage between shifts. Air in the pressure regulator valve will cause a buzzing noise when the valve tries to regulate pump pressure.

Excessively high fluid levels resulting from improper level checking procedures or by a transmission overheated from severe use can also cause aeration. As the planetary gears rotate in high fluid levels, air can be forced into the fluid. Aerated fluid can foam, overheat, and oxidize. All of these problems can interfere with normal valve, clutch, and servo operation. Foaming may be evident by fluid leakage from the transmission's vent.

4. Perform pressure tests; determine necessary action.

Electronically controlled transmissions have many similarities regardless of manufacturer. Transmissions equipped with electronic main line pressure control will generate maximum pressure when electricity is removed from the controls. This feature is meant to be a fail-safe design to protect the transmission. Most control devices of electronic transmissions are solenoids that control hydraulic pressure or volume to the transmission components. By varying the hydraulic main line pressure, the control module can control shift and converter clutch timing and feel. A solenoid operating at zero percent duty cycle is not activated at all. In this case, the transmission pressure would be at its highest point. If the solenoid is commanded at 100 percent duty cycle, it is fully open.

A pressure control solenoid operating at 100 percent duty cycle would cause low transmission pressure. The solenoid would be allowing a large volume of fluid to bleed out of the main line pressure circuit. The more volume that is allowed to bleed out, the lower main line pressure will be. It should be noted that solenoids can work in the opposite way (closing, as opposed to opening, when applied) so keep in mind zero percent is always when the solenoid is de-energized and 100 percent is fully energized. Most electrical circuits are designed to operate the solenoid at 60 percent duty cycle or less most of the time.

A common means of testing the control system and the ability of the transmission to generate pressure is to take a pressure reading when the throttle is at minimum opening (low throttle position sensor voltage) during idle. This should be the lowest pressure reading obtained. Disconnecting the pressure control solenoid or using a bidirectional scanner to command the solenoid off should generate the highest pressure. Refer to the manufacturer's service information for transmission-specific pressure tables to compare the actual readings to the specs. By comparing the actual readings to the specifications, the problem can be isolated as an electronic control issue or a mechanical transmission problem. Keep in mind that the transmission pump generates volume, the pressure regulator builds pressure, and the electronic pressure control (EPC) solenoid modulates pressure. The EPC solenoid will be largely responsible for pressure changes during operation.

In hydraullically shifted transmissions, many problems can be identified by conducting pressure tests. A pressure test is used to greatest value when the transmission shifts roughly or when the shift timing is wrong. Both of these problems may be caused by excessive line pressure.

During a road test, observe the starting pressures and the steadiness of the increases that occur with slight increases in load. The amount the pressure drops as the transmission shifts from one gear to another should also be noted. The pressure should not drop more than 15 psi (103 kPa) between shifts. Care should be taken with transmissions that use a cutback valve that drops pressure after the vehicle attains cruising speeds. Drops greater than 15 psi will occur and are normal. Also keep in mind that many cutback valves are known for sticking and causing low pressure under all conditions. Typically, when the fluid pressures are low, there is an internal leak, restricted filter, low oil pump output, or faulty pressure regulator valve. If the fluid pressure increased at the wrong time or the pressure was not high enough, sticking valves or leaking seals are indicated. If the pressure drop between shifts was greater than 15 psi, an internal leak at a servo or clutch seal is indicated.

To maximize the usefulness of a pressure test and be better able to identify specific problems, begin the test by measuring line pressure. Main line pressure should be checked in all gear ranges and at the three basic engine speeds. If the pressures in all operating gears are within specifications at slow idle, the pump and pressure regulators are working fine. If all pressures are low at slow idle, it is likely that there is a problem in the pump, pressure regulator, filter, or fluid level, or there is an internal pressure leak. To further identify the cause of the problem, check the pressure in the various gears while the engine is at fast idle. If the pressures at fast idle are within specifications, the cause of the problem is normally a worn oil pump; however, the problem may be an internal leak. Internal leaks typically are more evident in a particular gear range because that is when ATF is being sent to a particular device through a particular set of valves and passages. If any of these components leak, the pressure will drop when that gear is selected or when the transmission is operating in that gear.

Further diagnostics can be made by observing the pressure change when the engine is operating at wide-open throttle (WOT) in each gear range. A restricted oil filter will normally cause a gradual drop at higher engine speeds because the fluid cannot pass through the filter fast enough to meet the needs of the transmission and faster turning pump. If the fluid pressure did not change with the increase in engine speed, a stuck pressure regulator may still allow the pressure to build with an increase in engine speed, but it will not provide the necessary boost pressures. If the pressures are high at slow idle, the cause may be a faulty pressure regulator, throttle valve problem, or the transmission may be in fail-safe mode. Many electronic transmissions will raise line pressure, if a problem occurs, to protect the transmission and to alert the driver of a problem. This is called *fail safe* and will most often result in hard shifts.

If all of the pressures are low at WOT, pull on the TV cable or disconnect the vacuum hose leading to the vacuum modulator. If this causes the pressure to be in the normal range, the low pressure is caused by a faulty cable or there is a problem in the vacuum modulator. If the pressures stay below specifications, the most likely cause of the problem is the pump or the control system. If all pressures are high at WOT, compare the readings to those taken at slow idle. If they are high at slow idle and WOT, a faulty pressure regulator or throttle system is indicated. If the pressures are normal at slow idle and high at WOT, the throttle system is faulty. To verify that the low pressures are caused by a weak or worn oil pump, conduct a reverse stall test. If the pressures are low during this test but are normal during all other tests, a weak pump is indicated.

5. Perform stall tests; determine necessary action.

Before performing a stall test, insure that the transmission is warmed up and the fluid level is normal. A stall test puts high loads on the engine, transmission, drive train, and brakes. It should be performed with the driver fully in the vehicle and the area clear of obstructions and observers.

When performing a stall test, place the transmission in gear, hold the brake, and raise the engine's RPM. The stall speed is the highest engine speed achieved without turning the wheels. If the torque converter and transmission are functioning properly, the engine will reach a specific speed. If the tachometer indicates a speed above or below specification, a possible problem exists in the transmission or torque converter. If a torque converter is suspect it should be removed and the one-way clutch should be checked on the bench.

If the stall speed is below specifications, poor engine performance or a slipping stator clutch is indicated. If the stator's one-way clutch is not holding, ATF leaving the turbine of the converter works against the rotation of the impeller and slows down the engine. With both of these problems, the vehicle will exhibit poor acceleration, either because of lack of power from the engine or because there is no torque multiplication occurring in the converter. Do not perform the stall test for more than 15–20 seconds to prevent overheating the transmission.

If the stall speed is above specification, the bands or clutches in the transmission may be slipping and not holding properly. If the vehicle has poor acceleration but had good results from the stall tests, suspect a seized one-way clutch. Excessively hot ATF in the transmission is a good indication that the clutch is seized. However, other problems can cause these same symptoms, so be careful during diagnosis.

If a converter makes noise in gear, during the stall test put it in neutral; if the noises are no longer present, the source of the noise is probably the torque converter. The converter should be removed and bench tested for internal interference.

6. Perform torque converter clutch (lockup converter) mechanical/hydraulic system tests; determine necessary action.

Nearly all late-model transmissions are equipped with a lockup torque converter. Most of these lockup converters are controlled by the PCM. The PCM turns on the converter clutch solenoid, which opens a valve and allows fluid pressure to engage the clutch. Some manufacturers use a pulse-width modulated torque converter clutch to provide progressive lockup and unlock. The PCM in these applications uses RPM signal along with input and output shaft speeds to monitor and calculate the amount of converter clutch application it provides. If a converter or any other component of the transmission slips enough to exceed the criteria programmed into the computer software it will set a diagnostic trouble code (DTC) indicating a gear ratio problem. Many of the pulse-width modulated converters may be applied in any forward gear. Care should be taken during diagnostics because poor lockup clutch action can be caused by engine fuel and ignition systems, TCC solenoid, solenoid electrical and/or hydraulic control circuitry, controller module and/or inputs, the converter clutch itself, or other components within the torque converter. Many scan tools will allow you to modify, control, or override TCC application.

If during application of the TCC there is a concern with a shudder, there are several things to keep in mind. While the torque converter is often blamed for the problem, there are other possible causes that should be eliminated before removing the converter. A common cause of *perceived* TCC problems is an engine misfire that occurs when the converter locks up or as it is locking. Some ignition- and fuel-related misfires only happen under very specific conditions. It is also possible that a regular misfire cannot be felt until the converter is locked up. Another problem that causes TCC shudder during application is transmission fluid that is worn out. ATF contains friction modifiers that are critical to proper transmission operation. When these friction modifiers become ineffective from use, the transmission may demonstrate strange behavior: A shudder is common. In these cases, a simple fluid change may be all that is needed. Some manufacturers, in seeking that last ounce of gas mileage or emissions improvements, created shift schedules and

TCC strategies that were not successful in the real world. Be sure to check for technical service bulletins when you are diagnosing any kind of problem for updates to software or hard parts that might resolve your condition.

The timing of clutch application in conventional hydraulic transmissions is a common complaint in vehicles that have mechanical means of regulating transmission line pressure, such as a TV cable. If the adjustment of this cable is off it may cause the transmission to shift early or late. Transmissions that engage the clutch too early can cause the vehicle to experience a loss of power or even ping under light loads. The timing of the shift points and torque converter lockup are critical for proper operation and fuel efficiency.

2. Electronic Systems (14 Questions)

1. Road test the vehicle to verify electronic system problems based on driver's concern; determine necessary action.

Some transmissions are only partially electronically controlled; that is, only the engagement of shifting gears from third to fourth is electronically controlled. Other models feature electronic shifting into all gears plus electronic control of the TCC.

Critical to proper diagnosis of the electronic automatic transmission (EAT) and TCC control system is a road test. The road test should be conducted in the same way as one for a nonelectronic transmission, except that a scan tool should also be connected to the circuit to monitor engine and transmission operation. All pressure changes should be noted. The various computer inputs should also be monitored and the readings recorded for future reference. Some scan tools are capable of printing out a report. If the scanner does not have that function, data can be recorded while the customer's problem is occurring. Data leading up to the point of the problem will also be recorded for evaluation to determine the cause of the problem.

2. Perform pressure tests on transmissions equipped with electronic pressure control; determine necessary action.

Once again, remember that the transmission pump generates volume, the pressure regulator builds pressure, and the electronic pressure control (EPC) solenoid modulates pressure. The EPC solenoid will be largely responsible for pressure changes during operation. To expand that a little further, the PCM receives many inputs that it uses to control the EPC valve. Many of these are simply to make things such as power steering and air conditioning operate more smoothly and transparently. Let's take a look at one of these to illustrate a possible scenario. This is a brand-specific strategy but similar strategies exist in all vehicles. Imagine that the vehicle is at cruise speed and the driver turns on the air conditioning. With older hydraulically controlled transmissions the driver would be able to feel the A/C compressor engage. With EPC applications, the PCM can receive the A/C input and unlock the torque converter, letting it slip so that the A/C compressor clutch application is virtually seamless and undetectable in milliseconds. This type of strategy is common and important for you to be aware of when performing pressure tests. Since the PCM is using the EPC to modulate or adjust pressure and torque converter application you will see strange pressure changes occur.

Before you condemn a component make sure that what you are seeing is not a function of the design strategy of the vehicle. You can confirm this by looking at the EPC or torque converter electrical signals with a lab scope. If you rely strictly on a scanner to acquire your EPC pressure data, you will make diagnostic errors due to the latency (time delay) and the input averaging that will occur. If you need proof, simply connect a pressure gauge and see if your scanner tracks with the gauge. You will see that the data stream falls somewhere between real time and up to several seconds of delay time. As with other pressure control items, you will need to have a pressure chart or a shift schedule that tells you what the EPC pressure should be under certain situations. A question on an ASE test will give you this data if they want you to make a diagnostic decision. One more time, we will remind you that the default position for EPC is maximum pressure, zero percent duty cycle, low or no voltage applied to the EPC solenoid. Most failure strategies will cause hard shifts to protect against component failure.

3. Perform torque converter clutch (lockup converter) electronic system tests; determine necessary action.

All testing of TCC components should begin with a basic inspection of wires: Look for burned spots, bare wires, and damaged or pinched wires. Make sure the harness to the electronic control unit has a tight and clean connection. Also check the source voltage at the battery before beginning any detailed test on an electronic control system. If the voltage is not between 12 and 14.3 volts, the electronic system may not function properly.

On some early TCC-equipped vehicles, lockup was controlled hydraulically. A switch valve is controlled by two other valves. The first, the lockup valve, responds to governor pressure and prevents lockup at speeds below 40 mph (64 km/h). The second, the fail-safe valve, responds to throttle pressure and permits lockup in high gear only. Care should be taken during diagnostics because poor lockup clutch action can be caused by engine, electrical, clutch, or torque converter problems.

Before the lockup clutch is applied, the vehicle must be traveling at or above a certain speed. The vehicle speed sensor (VSS) sends this speed information to the computer. The converter should not be able to engage the lockup clutch when the engine is cold; an electronic coolant temperature sensor (ECT) provides the computer with information on temperature. On some early TCC applications, during sudden deceleration or acceleration, the lockup clutch should be disengaged. One of the sensors used by the PCM to control these driving modes is the throttle position sensor (TPS). A brake switch is used in some circuits to disengage the clutch when the brakes are applied. These key sensors—the VSS, ECT, TPS, and brake switch—should be inspected as part of your diagnosis.

4. Diagnose electronic transmission control systems using appropriate test equipment and service information (such as: shop manuals, technical service bulletins, schematics, etc.); determine necessary action.

Many switches are used as inputs or control devices for EATs. Among them, the transmission range sensor (TRS) is probably capable of causing more issues with several systems of the vehicle at one time than any other input. It can cause starting issues, a multitude of body control functions to fail, security system problems, and the list goes on. The one thing it can really confuse is the PCM that is trying to run the transmission. The TRS can be digital or analog. The digital type uses an on/off signal for each shift detent position and it is pretty easy to diagnose with a scan tool looking for each input as the shifter is moved. The analog type is more complex because the PCM is looking for a specific voltage at each position throughout shifter travel. For example, the drive position is 2.5 volts and second-gear position is 2.1 volts. What is 2.3 volts? This is where the PCM can be confused and make bad choices. Check and adjust an analog TRS with a scan tool. This way you can be sure that the PCM is getting the correct signal.

Another important thing to keep in mind is that many vehicles that use pressure switches offer you and the PCM a way to confirm that, if the TRS is signaling for a particular gear, the hydraulic action needed to engage that gear has occurred. This is evident if you are seeing the switch close on the scan tool. The information and specs necessary to adjust and diagnose the TRS will have to be gathered from your information system or the appropriate manual.

Pressure switches either complete or open an electrical circuit. They are inputs to the computer and may be used for many purposes, but primarily tell the PCM digitally that the transmission has indeed completed a shift and is in a particular gear. Normally, open switches will have no continuity across the terminals until oil pressure is applied to them. Your scan tool will tell you if the switches are functioning.

The TPS is a potentiometer. A potentiometer is a three-wire sensor that consists of a voltage reference wire, a signal wire, and a signal return (ground) wire. The TPS sends a voltage signal that corresponds with the actual throttle position to the PCM. Based on that voltage signal, the PCM knows the exact location of the throttle. The TPS signal is very low voltage, usually 0.5–1.0 V DC to the computer when the throttle is closed and increases the voltage to around 4–5 V DC as the throttle is opened. The TPS signal can affect main line pressure and shift schedule more than any other input. Testing a TPS should be performed with a lab scope so that you can see any kind of voltage signal dropouts or spikes in the potentiometer as the throttle is opened.

The VSS is another critical transmission input to the PCM that can cause problems throughout the vehicle. The VSS creates a digital pulse train or AC sine wave, depending on the type used and when viewed on a lab scope, for virtually all vehicles. This signal is passed along to body control modules to run automatic door locks, windshield wipers, and cruise control among the more common items. It is also used to drive the speedometer. The PCM uses the VSS signal as an input to run shift schedules and many engine operations such as OBD-II monitors and exhaust gas recirculation (EGR) valve control.

Temperature sensors are mainly of two varieties since their inclusion on electronic transmissions. Early and some current designs infer temperature of the transmission from the engine's temperature sensor, and if equipped with its own controller, the transmission shares this information with the engine controller, usually via a data bus between the two controllers. EATs are equipped with temperature sensors located in the reservoir or outlet cooler line. Temperature sensors are designed to change resistance with changes in temperature. In most transmission fluid temperature sensors, the resistance of the sensor will decrease as the temperature of the fluid increases. This will cause the voltage from the PCM to drop as the temperature heats up. A cold temperature is indicated by a high signal voltage, while a hot temperature is indicated by a low voltage. It will be necessary to refer to manufacturer data to test the sensor properly. Since these are fairly slow-moving signals it is safe to perform preliminary testing by comparing actual temperature to the information provided by the scan tool. If a problem is found, it will be necessary to look for resistance occurring within the wiring. The problem could also be a faulty sensor.

Turbine speed and output shaft speed sensors are used to monitor the shaft speeds within the transmission. The transmission monitors these sensors very closely while the shifting is taking place. They allow for complex shift schedules that allow the engine to operate at the appropriate torque output for almost any driving condition. They are also used for comparative diagnostic operations within the PCM to determine torque converter and gear train slippage. Their signals are very similar to the VSS signal and can best be monitored with a lab scope in a real-time driving situation. The scan tool is also an excellent way to determine the perceived inputs from each PCM.

5. Verify proper operation of charging system; check battery, connections, vehicle power, and grounds.

The extensive use of electronics in vehicles necessitates that all electrical systems are operating correctly. Poor connections can cause voltage drops that will cause drivability issues and even damage. A poor ground can cause damage to any rotating, conductive part of the vehicle including the pumps inside the transmission. Usually, starting system problems that extend beyond the electrical part of the starting system involve the flexplate on the back of the engine and may damage the gear teeth of the flexplate. It is important in any area of automotive repair to have a good understanding of the electrical system and basic electrical theory. Be sure that you are familiar with performing voltage drops, Ohm's Law, and starting/charging system tests before taking the A2 test.

6. Differentiate between engine performance, or other vehicle systems, and transmission/transaxle-related problems; determine necessary action.

Correct up- and down-shift timing of the automatic transmission enhances the performance of the engine. When engine emission control systems were incorporated, engine power became marginal. Misadjusted or malfunctioning throttle pressure and governor pressure controls, causing early transmission up-shifts and/or late or no part throttle or WOT down-shifts, can seriously affect vehicle acceleration and overall performance. Also, the advent of TCC systems design strategies often has the lockup clutch apply immediately after shifting into direct or even second gear under certain operating conditions. Thus, early shift points can cause a host of problems, ranging from engine detonation to vibration on acceleration. Shift timing specifications for mechanically operated shift control systems that list shift points versus mph for various final drive ratios and tire size combinations can sometimes be found in repair manuals. It is important to check the shift points when road testing the vehicle. The engine controller often operates the torque converter lockup clutch. There are times when an updated programmable read-only memory (PROM) chip or controller is available to correct (raise) the lockup clutch apply point. Most late-model vehicles use an electronically erasable PROM which can be reprogrammed with a scan tool.

Check TSBs for symptoms that relate to your concern. If no problems are found related to the transmission, and vehicle performance is unsatisfactory, the engine should be tested for low output.

A common scenario today often manifests itself in the form of a vehicle intermittently and momentarily bucking/surging or shuddering sharply until load conditions change. This occurs when vehicle load becomes greater than what the TCC can transmit, causing a break loose/grab condition between the lockup clutch piston and the torque converter cover. A momentary surge or buck may also be the result of a fuel or ignition system malfunction causing a load-related misfire or series of misfires that cause power not to be delivered smoothly from the crankshaft to the lockup clutch. Due to the similarity of the symptoms, it is very difficult to differentiate whether the problem stems from an engine misfire or a lockup clutch shudder. Ported fuel injection systems complicate matters as individual cylinder fuel system problems feel very much like what an individual ignition misfire feels like.

One widely accepted method of determining whether a problem exists in the engine or transmission is to use a lab scope to monitor one or more upstream oxygen sensor waveforms while the problem is occurring. If the sensor is functioning properly, it will quickly indicate if a cylinder misfire occurred and whether the misfire most likely occurred from a lean cylinder charge, insufficient spark intensity or duration, or excessive EGR flow to a cylinder(s). With the advent of OBD-II emission control systems, many transmission control strategies eliminate certain functions such as TCC operation when an engine misfire code is present or pending in the controller. This always makes it a necessary practice to check for the presence of any engine codes stored in both the manufacturer and OBD-II generic side of the controller. If testing shows no signs of an engine problem, a scanner should be used to test the slippage, measured in RPM, that exists between the engine and the turbine shaft. Observe this value while the symptom is occurring. If there is excessive slippage compared to normal slip, the RPM problem is with the transmission/ torque converter.

7. Diagnose shift quality concerns resulting from problems in the electronic transmission control system; determine necessary action.

While there are no blanket statements that cover all fail-safe diagnostic strategies involving all EATs, most systems fail in one or more of these three categories. The first category is when the processor places the transmission in a fixed gear, a no shift condition. This condition is often called *default mode, limp-in mode*, or *hardware limited operation strategy* (HLOS). If a transmission is powered by a relay, this type of failure usually results in shutting down complete electrical power to the transmission via the relay. This condition can be brought about by: slippage, indicated by the input and output speed sensors ratio errors while in gear or during a shift; control system actuator opens or shorts; and no or erratic, input or output shaft speed sensor signals. If the problem is a hard fault, a code should be stored. Retrieve the code, understand the code-setting criteria, and follow the trouble chart for hard code diagnosis. If the code is a soft fault, get the transmission to operating temperature (208°F minimum). While test driving, observe a scanner for slippage indication by recording speed sensor ratios for the indicated range or shift. If a pressure tap is available, often a pressure test will verify if the problem is a circuit leak.

Observing the amperage waveform of the shift solenoid while operating it with a scanner or a shift controller box will verify the electrical and mechanical integrity of the solenoid. Intermittent speed sensor signals can often be traced to poor connections. An ohmmeter attached across the disconnected circuit at the processor end of the harness, while performing a wiggle test, will often find the faulty connection. Ultimately, the goal is to determine whether or not the unit must be pulled out of the vehicle to be repaired.

In the second category of fail-safe operation, the computer recognizes a problem but still controls the shift pattern. If equipped with electronic main line pressure control, the shifts will be firmer than normal and shifts may not include all gears. A pressure test will determine if the transmission is operating in this mode. While operating in this mode, line pressure will be constant and high (no line pressure rise with throttle movement). The solenoid control circuit can be observed with an amp probe to verify solenoid winding and core movement integrity while performing this test. Manufacturers dub this category *failure mode effects management* (FMEM), *shift strategy abort*, or *default action*.

Typically, a code is set and the gear having the shift problem is eliminated for the current key cycle or until the code is erased from the controller's memory. These problems are often intermittent and often require high transmission temperatures to re-create the conditions necessary to reveal the problem. Test driving with a scanner and a pressure gauge attached to the circuit will often determine whether the problem is electrical or mechanical.

The third category is one that warns the driver with a flashing OD, S4, or Hold light. Often, this mode will only alter shift timing, while in some situations it will eliminate the gear in which the shift is occurring. A trouble code will be stored, and the light will continue to flash as long as the problem continues. These problems are most often electrical.

B. In-Vehicle Transmission/Transaxle Maintenance and Repair (12 Questions)

1. Inspect, adjust, and replace manual valve shift linkage, transmission range sensor/switch, and park/neutral position switch (inhibitor/neutral safety switch).

Many computer-controlled vehicles use a multiposition switch called a transmission range sensor (TRS) to provide an input to the PCM. Refer to Task A.2.4 for more information on the TRS. Older vehicles that have neutral safety switches and late-model vehicles with TRSs can all have situations where they will not allow cranking for starting or turn on back-up lamps due to misadjusted shift linkage or a misadjusted sensor/switch. A worn or misadjusted gear selection linkage may affect transmission operation. The transmission's manual shift valve must completely engage the selected gear. Partial manual shift valve engagement may not allow the proper amount of fluid pressure to reach the rest of the valve body, resulting in poor gear engagement, slipping, and excessive wear. Keep in mind that the internal manual shift linkage that connects to the valve body is more likely to cause these types of problems. Most late-model vehicles have cable-operated shift with an adjustment at one end. Other than a failure to engage park fully, most customer complaints concerning the external linkage are related to the shift pointer being off from the gear; that typically is an adjustment separate from the gear shift linkage itself.

2. Inspect, adjust, and replace cables or linkages for throttle valve (TV), kick down, and accelerator pedal.

The throttle valve cable connects the movement of the throttle to the TV in the transmission's valve body. On some transmissions, the throttle linkage may control both the downshift valve and the throttle valve. Others use a vacuum modulator to control the TV and throttle linkage to control the downshift valve. The action of the throttle valve produces throttle pressure. Throttle pressure is used as an indication of engine load and influences the speed at which automatic shifts will take place.

A misadjusted TV linkage may also result in a throttle pressure that is too low in relation to the amount the throttle plates are open, causing early up-shifts. Throttle pressure that is too high can cause harsh and delayed up-shifts, and part throttle and WOT downshifts will occur earlier than normal. Very small changes in throttle cable adjustment can make a big difference in shift timing and feel.

Most late-model transmissions are not equipped with downshift linkages. Some transmissions use an electric kickdown switch, typically located at the upper post of the throttle pedal. Movement of the throttle pedal to the wide-open position signals to the transmission that the driver desires a forced downshift. To check the switch, fully depress the throttle pedal and listen for a click that should be heard just before the pedal reaches its travel stop.

If the click is not heard, loosen the locknut and extend the switch until the pedal lever makes contact with the switch. If the pedal contacts the switch too early, the transmission may downshift too early. If you hear the click but the transmission still does not downshift, use an ohmmeter to check the switch. An open switch will prevent forced downshifts, whereas a shorted switch can cause up-shift problems.

3. Inspect and replace external seals and gaskets.

Leaks are often caused by defective gaskets or seals. Common sources for leaks are the oil pan seal, rear cover, final drive cover (on transaxles), extension housings, speedometer drive gear assembly, and electrical switches mounted into the housing. The housing itself may have a porosity problem, allowing fluid to seep through the metal.

A common cause of fluid leakage is the seal of the oil pan to the transmission housing. If there are signs of leakage around the rim of the pan, retorquing the pan bolts may correct the problem unless the pan is sealed with formable gasket sealer (silicone). If tightening the pan does not correct the problem, the pan must be removed and a new gasket installed. Make sure the sealing surface of the pan's rim is flat and capable of providing a seal before reinstalling it.

An oil leak at the speedometer cable can be corrected by replacing the housing o-ring, driven gear shaft seal, and the driven gear if necessary. While replacing the seal, inspect the drive gear for chips and missing teeth. Always lubricate the o-ring and gear prior to installation.

4. Inspect and replace drive shaft yoke, drive axle joints, bushings, and seals.

An oil leak stemming from the mating surface of the extension housing and the transmission case may be caused by loose bolts. To correct this problem, tighten the bolts to the specified torque. Also check for signs of leakage at the rear of the extension housing. Fluid leaks from the seal of the extension housing can be corrected with the transmission in the vehicle. Often, the cause for the leaks is a worn extension housing bushing that supports the sliding yoke of the drive shaft. When the drive shaft is installed, the axial (up-and-down) or lateral (side-to-side) movement should be minimal. Specifications for clearance are given, but a more practical measurement of wear is with the drive shaft removed. Compare the oil groove depth of the bushing to a new one, which also pinpoints wear area. A bent shaft or a damaged spline can cause both abnormal wear and noise. If the clearance is satisfactory, a new oil seal will correct the leak. If the clearance is excessive, the repair requires that a new bushing be installed. If the leak persists, the transmission vent should be checked for blockage.

5. Check condition of engine cooling system; inspect transmission lines and fittings.

Since the transmission cooler is often a part of the radiator, the engine's cooling system is the key to efficient transmission fluid cooling. If anything affects engine cooling, it will also affect ATF cooling. The engine's cooling system should be carefully inspected whenever there is evidence of ATF overheating or a transmission cooling problem. If the problem is the transmission cooler, examine it for signs of leakage. The most common type breech of fluid in the cooler system is ATF leaking into the engine coolant. This is because of the operating pressure differences that exist between the two systems. This is evident by the presence of oil floating to the top of the cooling system and becoming visible through the fill cap or an oily film floating on top in the overflow bottle. In cases that are left unattended, the rubber components of the cooling system (hoses, for example) can swell and feel oily to the touch. If the leak becomes bad enough, it usually becomes a two-way leak, now leaking coolant into the transmission. This is evident by the ATF having a milky appearance or possibly the dipstick having rust deposits. A leaking transmission cooler core can be verified with a leak test.

Check the condition of the cooler lines from their beginning to their end. A line that has been accidentally damaged while the transmission has been serviced will reduce oil flow through the cooler and shorten the life of the transmission. If the steel cooler lines need to be replaced, use only the appropriate steel tubing. Never use copper or aluminum tubing to replace steel tubing. The steel tubing can be double flared and installed with the correct fittings.

6. Inspect valve body mating surfaces, bores, valves, springs, sleeves, retainers, brackets, check balls, screens, spacers, and gaskets; replace as necessary.

If the pressure test indicates there is a problem associated with the valves in the valve body, a thorough disassembly, a cleaning in fresh solvent, a careful inspection, and the freeing up and polishing of the valves may correct the problem. Sticking valves and sluggish valve movements are caused by poor maintenance, the use of the wrong type of fluid, and/or overheating the transmission. The valve body of most transmissions can be serviced when the transmission is in the vehicle, but is typically serviced when the transmission has been removed for other repairs.

After all of the valves and springs have been removed from the valve body, soak the valve body and separator plates in mineral spirits for a few minutes. Some rebuild shops soak the valve body and its associated parts in carburetor cleaner, and then wash off the parts with water. Thoroughly clean all parts and make sure all passages within the valve body are clear and free of debris. Carefully blow-dry each part individually with dry, compressed air.

Check the separator plate for scratches or other damage. Scratches or score marks can cause oil to bypass correct oil passages and result in system malfunction. If the plate is defective in any way, it must be replaced. Check the oil passages in the upper and lower valve bodies for varnish deposits, scratches, or other damage that could restrict the movement of the valves. Check all of the threaded and related bolts and screw for damaged threads, and replace as needed.

Examine each valve for nicks, burrs, and scratches. Make sure that each valve fits properly into its respective bore. To do this, hold the valve body vertically and install an unlubricated valve into its bore. Let the valve fall of its own weight into the valve body until the valve stops. Placing your finger over the valve bore, turn the valve body over. The valve should again drop by its own weight. If the valve moves freely under these conditions, it will operate freely with fluid pressure.

If a steel spool valve cannot move freely within its bore, it may have small burrs or nicks. These flaws can and should be removed. To do this, never use sandpaper or a file; rather, use products, such as an Arkansas stone or crocus cloth, which are designed to polish the surface without removing metal from the valve. Sandpaper and emery cloth will remove metal, as well as scratch and leave a rough surface.

After polishing, the valve must be thoroughly cleaned to remove all of the cleaning and abrasive materials. After the valve has been cleaned, it should be tested in its bore again. If the valve cannot be cleaned well enough to move freely in its bore, the valve body should be replaced. If the valve body has been severely contaminated with metal particles it should be replaced, especially if it is aluminum. Replacement is necessary for several reasons. Although some parts are available separately for valve bodies, spool valves usually are not. Another reason is that aluminum valve body bore-to-spool clearances are much tighter than older cast-iron bodies, making it easy to create an oversized or out of round bore using conventional techniques for bore repair. Also, techniques used in the past to correct a scratched valve spool land do not work well with aluminum or *anodized* aluminum valve spools. Individual valves are lapped to a particular valve body, and, therefore, if any parts need to be replaced, the entire valve body must be replaced.

Although it is desirable to have the valves move freely in their bores, excessive wear is also a problem. With the advent of EATs and the use of pulse-width modulated (PWM) solenoid control, certain valve spools are stroked at very high frequencies to modulate pressures and vary device apply rates. These very high operating frequencies require lightweight aluminum valve spools to maintain accurate control of fluid flow rates. Along with these control systems has come a new set of problems: valve spool and/or bore wear. Several methods to assess and correct the symptoms of this bore wear are offered by different manufacturers that do not require complete valve body replacement.

7. Check and adjust valve body bolt torque.

Before beginning to reassemble a valve body, check the new valve body gasket to make sure it is the correct one by laying it over the separator plate and holding it up to the light. No oil holes should be blocked. Then install the bolts to hold the valve body sections together and the valve body to the case. Tighten the bolts to the torque specifications to prevent warpage and possible leaks. Overtorquing can also cause the bores to distort, which would not allow the valves to move freely once the valve body is tightened to the transmission case.

8. Inspect accumulator and servo bores, pistons, seals, pins/pin bores, springs, and retainers; repair or replace as necessary; adjust bands.

On some transmissions, the servo and accumulator assemblies are serviceable with the transmission in the vehicle. Others require the complete disassembly of the transmission. A technician should check with the manufacturer's recommendation to determine which action is needed.

On those units equipped with servos or accumulators serviced externally, carefully inspect the cover area to determine the exact cause of the leakage. Do this before cleaning the area around the seal. Look at the path of the fluid leakage and identify other possible sources. These sources could be worn gaskets, loose bolts, cracked housings, or loose line connections.

When removing the seal, inspect the sealing surface, or lips, before washing. Look for unusual wear, warping, cuts and gouges, or particles embedded in the seal.

Band servos and accumulators are basically pistons with seals in a bore that are held in position by springs and retaining snap-rings. Remove the retaining rings and pull the assembly from the bore for cleaning. Check the condition of the piston and springs. Cast-iron seal rings may not need replacement, but rubber and elastomer seals should always be replaced.

A servo's piston, spring, and piston/guide pin should be cleaned and dried. Check the servo piston for cracks, burrs, scores, and wear. Servo pistons may be made of either aluminum or steel. Aluminum pistons should be carefully checked for cracks and piston/guide pin bore wear. On both aluminum and steel the seal groove should be free of nicks or any imperfections that pinch or bind the seal. Clean up these problems with a scraper or small file. The servo bore should be checked for severe scratches and localized wear areas. Any of these conditions can cause pressure loss and/or piston binding, resulting in shift problems and/or low line pressure problems.

Minor bore scratches can be polished out, and there are bore sleeves available from manufacturers to salvage a transmission case with severely worn servo bores, and oversized piston/guide pins available to repair piston wear problems. A side clearance of 0.003–0.005 inch (0.076–0.127 mm) is required. Bands in the transmission may require adjustment. Many are self-adjusting and others may require loosening a lock nut and tightening to a specified torque, then a back-up rotation by a specified number of turns, and finally tightening the locknut. This procedure tightens the band around the braking drums, and the loosening of the adjustment stem provides the important maintained clearance.

9. Inspect, test, adjust, repair, or replace electrical/electronic components and circuits including computers, solenoids, sensors, relays, terminals, connectors, switches, and harnesses.

Some electronic transmissions are only partially controlled. Only the engagement of the converter and third to fourth shifting is electronically controlled. Other models feature electronic shifting into all gears, plus electronic control of the TCC.

The controls of an EAT direct the hydraulic flow through the use of solenoid valves. When it is used to control TCC operation, the solenoid opens a hydraulic circuit to the TCC spool valve, causing the spool valve to move and a regulated (reduced) form of line pressure to apply the converter clutch. This clutch apply oil circuit has been given numerous names by manufacturers. Electronically controlled shifting is accomplished in much the same way. Shifting occurs when a solenoid is either turned on or turned off. At least two shift solenoids are incorporated into the system, and shifting takes place by controlling the solenoids. The desired gear is put into operation through a combination of on and off or constantly duty-cycled solenoids, with elevated ground potential to modulate flow and guard against overheating the solenoid.

Several sensors and switches are used to inform the control computer about the current operating conditions. Most of these sensors are also used to calibrate engine performance. The computer then determines the appropriate shift time for maximum efficiency and best feel. The shift solenoids are controlled by the computer, which either supplies power to the solenoids or supplies a ground circuit. The techniques for diagnosing electronic transmissions are basically the same techniques used to diagnose TCC systems.

Although EATs are rather reliable, they have introduced new problems for the automatic transmission technician. Some of the common problems that affect shift timing and quality, as well as the timing and quality of TCC engagement, are wrong battery voltage, a blown fuse, poor connections, a defective TPS or VSS, defective solenoids, crossed wires to the solenoid or sensors, corrosion at an electrical terminal, or the faulty installation of some accessory, such as a rear entertainment system.

Improper shift points can be caused by electrical circuit problems, faulty electrical components, or bad connectors, as well as a defective governor or governor drive gear assembly. Some EATs do not rely on the hydraulic signals from a governor; rather they rely on the electrical signals from electrical sensors to control a PWM solenoid that converts line pressure into a pressure linear to road speed, which controls the timing of the shift valves.

Computer-controlled transmissions often start off in the wrong gear. This can happen due to either internal transmission problems or external control system problems. Internal transmission problems can be faulty solenoids or stuck valves. External problems can be the result of a complete loss of power or ground to the control circuit or a fail-safe protection strategy initiated by the computer to protect itself or the transmission from an observed problem. Sometimes, the default gear is simply the gear that is applied when all the shift solenoids are off, usually second or third gear and reverse, to allow limp-in operation to continue driving. If ECT is equipped with electronic pressure control, main line pressure will be defaulted to a fixed high level to prevent apply clutch or band burn-up.

A visual inspection of the transmission and the electrical system should include a careful check of all electrical wires and connectors for damage, looseness, and corrosion. Loose connections, even when clean, usually only make intermittent contact. They will also corrode and collect foreign material, which can prevent contact altogether. Control devices, such as solenoids, are used in higher current applications.

The electrical portion of these control circuits is best checked with the circuit powered, using the voltage-drop method and a DVOM or oscilloscope to see what portion of the circuit has the poor connection. Some systems offer bi-directional testing with a scan tool that often allows control of the solenoid duty cycle. This presents an excellent opportunity for volt-drop testing under maximum current demand. Low current input signals, such as speed sensors, often suffer from intermittent connections. A simple but effective method of checking these is with an ohmmeter attached across the disconnected sensor leads at the controller, while a wiggle test is performed on the circuit and its connections. An erratic meter reading indicates a faulty connection at the area being wiggled. Check all ground straps to the frame or engine block. This part of your inspection is especially important for electronically controlled transmissions that have a lockup torque converter. Check the fuse or fuses to the control module. To accurately check a fuse, either test it for continuity with an ohmmeter or check each side of the fuse for open power when the circuit is activated.

▪ 10. Inspect, replace, and align power train mounts.

Many shifting and vibration problems can be caused by worn, loose, or broken engine and transmission mounts. Visually inspect the mounts for looseness and cracks. To get a better look at the condition of the mounts, pull up and push down on the transaxle case while watching the mount. If the mount's rubber separates from the metal plate or if the case moves up but not down, replace the mount. If there is movement between the metal plate and its attaching point on the frame, tighten the attaching bolts to the appropriate torque.

Then, from the driver's seat, apply the foot brake and start the engine. Put the transmission into a gear and gradually increase the engine speed to about 1,500–2,000 rpm. Watch the torque reaction of the engine on its mounts. If the engine's reaction to the torque appears to be excessive, broken, or worn, drive train mounts may be the cause. Late-model vehicles equipped with hydraulic engine mounts may show no signs of being visibly torn, but can be structurally torn inside. This type of mount will allow more movement than a conventional mount when subjected to engine torque-type testing, even when it is performing normally.

Experience proves to be the best judge of a failed mount of this type when subjected to engine torque-type testing. If it is necessary to replace the transaxle mount, make sure you follow the manufacturer's recommendations for maintaining the alignment of the driveline. Failure to do this may result in poor gear shifting, vibrations, and/or broken cables. Some manufacturers recommend that a holding fixture or special bolt be used to keep the unit in its proper location.

When removing the transaxle mount, begin by disconnecting the battery's negative cable. Disconnect any electrical connectors that may be located around the mount. It may be necessary to move some accessories, such as the horn, in order to service the mount without damaging some other assembly. Be sure to label any wires you remove to facilitate reassembly.

Install the engine support fixture and attach it to an engine hoist. Lift the engine just enough to take the pressure off of the mounts. Remove the bolts attaching the transaxle mount to the frame and the mounting bracket, and then remove the mount.

To install the new mount, position the transaxle mount in its correct location on the frame and tighten its attaching bolts to the proper torque. Install the bolts that attach the mount to the transaxle bracket. Prior to tightening these bolts, check the alignment of the mount. Once you have confirmed that the alignment is correct, tighten all loosened bolts to their specified torque. Remove the engine hoist fixture from the engine, and reinstall all accessories and wires that may have been removed earlier.

▪ 11. Replace fluid and filter(s); verify proper fluid level and type.

During a fluid and filter replacement, the transmission oil pan will need to be removed. A large catch pan or funnel will be needed to catch the fluid as many transmission oil pans do not have a drain plug. Once the pan has been removed, the filter can then usually be seen. Although rare, some vehicles will have a spin-on cartridge filter similar to an engine oil filter. With the pan removed, check for any debris. A slight amount of metal stuck to the magnet will be normal. Excessive metal should be cause for alarm and the customer should be informed. Large metal pieces usually indicate a serious problem. A black film in the bottom of the pan typically comes from the clutches. An excessive amount of clutch material indicates a worn transmission.

If all inspections pass, clean all gasket mating surfaces, and make sure there are no raised areas on the pan due to the oil pan fasteners. Install the pan using the proper fastener torque, as excessive torque will distort the pan and inadequate torque will leave the pan loose, both of which will cause a leak. With the pan installed, refer to the service information for the proper transmission fluid type as many manufacturers have their own specific type. Fill the transmission with the specified amount of the proper fluid. Many manufacturers have their own specified type of fluid. Newer versions of these fluids are introduced and are usually backward compatible so that the older fluid will not have to be kept on hand. While there are many newer vehicles with various methods of checking transmission fluid, the most common method is with the use of a dipstick. Most vehicles specify checking the fluid level with the engine running. Fluid is to be checked with the transmission in park on most vehicles, while others will need to be in neutral. Fill the transmission to the cold line (if one is present). Start the vehicle and allow it to run for a few minutes to get the fluid hot. Check the fluid level after the fluid is hot and verify that the fluid in now to the hot level and add if necessary.

A simple fluid and filter change will not replace all of the transmission fluid. The fluid in the torque converter will not drain with the fluid in the pan. New equipment has been introduced that will flush the transmission and replace the fluid that is in the torque converter as well as the fluid in various lines and passages that would not normally drain into the pan. Many of these transmission-flushing machines will connect inline with the transmission cooler. The disadvantage of using one of these machines is that the pan is not removed and the filter is not replaced.

C. Off-Vehicle Transmission/Transaxle Repair (13 Questions)

1. Removal and Installation (4 Questions)

1. Remove and replace transmission/transaxle; inspect engine core plugs, rear crankshaft seal, transmission dowel pins, dowel pin holes, and mating surfaces.

The first step in removing the transmission is to disconnect the battery. Removing the transmission from an RWD vehicle is generally more straightforward than removing one from an FWD model, as there is typically one cross member, one drive shaft, and easy access to cables, wiring, cooler lines, and bell housing bolts. Transmissions in FWD cars, because of their limited space, can be more difficult to remove as you may need to disassemble or remove large assemblies such as the engine cradle, suspension components, brake components, splash shields, or other pieces that would not usually affect RWD transmission removal.

On RWD vehicles, raise the vehicle and drain the transmission fluid. Mark the drive shaft at the rear axle before disconnecting it to avoid run-out-related vibrations. Remove the drive shaft.

On FWD vehicles, attach a support fixture to the engine, raise the vehicle, and drain the fluid. Remove the front wheels and follow the service manual to remove the front axles.

Disconnect manual linkages, vacuum hoses, electrical connections, speedometer drives, and control cables. The inspection cover between the transmission and the engine should be removed next. Mark the position of the converter to the flexplate to help maintain balance or runout. It will be necessary to rotate the crankshaft to remove the converter bolts. This can be done by using a long ratchet and socket on the crankshaft bolt or by using a flywheel turning tool if space permits.

Position a transmission jack before removing any cross member or bell housing bolts. The use of a transmission jack also allows for easier access to parts hidden by cross members or hidden in the space between the transmission and the vehicle's floor pan. While supporting the transmission with a transmission jack, remove the cross member or transmission mounts. Remove the starter and bell housing bolts.

Now pull the transmission away from the engine. It may be necessary to use a pry bar between the transmission and engine block to separate the two units. Make sure the converter comes out with the transmission. This prevents bending the input shaft, damaging the oil pump, or distorting the drive hub. After separating the transmission from the engine, retain the torque converter in the bell housing. This can be done simply by bolting a small combination wrench to a bell housing bolt hole across the outer edge of the converter. When the transmission is removed, it is an opportune time to check the engine's rear crankshaft seal. Most late-model rear main engine seals are one piece and accessed by only engine or transaxle/transmission removal.

Be sure to check all bell housing bolt holes and dowel pins. Cracks around the bolt holes indicate that the case bolts were tightened with the case out of alignment with the engine block. The case should be replaced if any cracks are present. A transmission case is very thin, and welding may distort the case. It is not possible to determine if a repair will hold.

If any of the bolts that were removed during disassembly have aluminum on the threads, the thread bore is damaged and should be repaired. Thread repair entails installing a threaded insert, which serves as new threads for the bolt, or retapping the bore. After the threads have been repaired, make sure you thoroughly clean the case.

2. Inspect converter flex (drive) plate, converter attaching bolts, converter pilot, converter pump drive surfaces, converter end-play, and crankshaft pilot bore.

With the transmission removed, inspect the torque converter flex or drive plate. Check for any cracks that may be evident. A cracked flexplate can often make a knocking noise. If the flexplate contains the starter ring gear, check the teeth for any damage and replace if damage is evident. Check the flexplate mounting holes; any elongation indicates that the torque converter mounting bolts were loose.

Before removing the torque converter, inspect the pilot that fits into the crankshaft on the front of the converter. This pilot centers the converter with the crankshaft. Any wear on the pilot will result in an off-center torque converter and possible vibration. Any wear on the pilot should result in converter replacement. It is also vital to check for wear in the pilot bore in the end of the crankshaft as well.

Before removing the converter, it may be wise to measure how deep the converter sets into the bell housing so that upon installation you can be sure that the converter is installed all the way. With the measurement taken and the converter removed, check the pump drive hub for any wear. The pump drive may have two notches or grooves used to drive the transmission oil pump. Any damage to this hub should result in converter replacement. Also check the area that will mate with the front pump seal for damage. Any scoring or grooves that would result in seal damage and leaks should result in converter replacement.

With the converter removed, check the converter end-play. This involves using a special adapter that expands to tightly grab the stator inside the torque converter. A dial indicator is mounted to the adapter and the stylus is placed on the converter housing. The adapter is pulled upward measuring the amount of movement of the stator. Excessive end-play indicates that internal thrust washers have worn. In today's modern welded torque converters, the best fix is torque converter replacement. A simpler, but less accurate, method of checking torque converter end-play is by using a pair of snap-ring pliers to grab the stator. Lift the pliers to approximate how much stator movement is present. The one-way clutch of the stator can also be tested at this time. The clutch should turn freely in one direction and lock in the other.

3. Inspect, test, flush, or replace transmission oil cooler.

Vehicles equipped with an automatic transmission can have an internal or external transmission cooler, or both. The basic operation of either type of cooler is that of a heat exchanger. Heat from the fluid is transferred to something else, such as liquid or air. Hot ATF is sent from the transmission to the cooler, where its heat is removed, and then the cooled ATF returns to the transmission.

Internal coolers are located inside the engine's radiator. Heated ATF travels from the torque converter to a connection at the radiator. Inside the radiator is a small internal cooler, which is sealed from the liquid in the radiator. ATF flows through this cooler and its heat is transferred to the liquid in the radiator. The ATF then flows out of a radiator connection, back to the transmission.

External coolers are mounted outside the engine's radiator, normally just in front of it. Air flowing through the cooler removes heat from the fluid before it is returned to the transmission.

A faulty transmission oil cooler can cause damage to a new transmission. Check the cooler for any leaks that could cause low fluid levels and lack of transmission cooling. Any obstructions in airflow can cause a lack of cooling, which can severely damage a transmission. Clean out any debris that can block airflow. Cooling fins that are severely bent or damaged should be straightened or the cooler replaced.

Debris can also get trapped inside the transmission oil cooler. This debris comes from the damaged internal components of the transmission. If this debris is not flushed out, it can enter the new transmission and cause damage or improper operation. Any time that the transmission is replaced, the cooler should be flushed with an approved solvent. Back-flushing is usually suggested. If back-flushing does not clear the blockage in the cooler, then the cooler will have to be replaced.

If a clogged cooler is suspected, it can be checked for pressure drop and flow. To check for pressure drop, install an oil pressure gauge before and after the oil cooler. Start the vehicle and observe the gauges. There should be very little pressure difference in the two gauges. A blockage in the cooler will result in a large pressure difference. To check for fluid flow, remove the cooler outlet hose and route it into a container to catch the fluid. Start the vehicle and observe the amount of fluid that enters the can in 30 seconds. If specs are available, compare the readings; if not, the pump should generally flow at least a quart in 20 seconds.

2. *Disassembly and Assembly (5 questions)*

1. Disassemble, clean, and inspect transmission case, subassemblies, mating surfaces, and thread condition.

Before disassembling the automatic transmission, care should be taken to clean away any dirt, undercoating, grease, or road grime on the outside of the case. This ensures that dirt will not enter the transmission during disassembly. Once the transmission is clean outside, you may begin the disassembly. When cleaning automatic transmission parts, avoid the use of solvents, degreasers, and detergents that can decompose the friction composites used in a transmission. Use compressed air to dry components; do not wipe down parts with a rag. The lint from a rag can easily destroy a transmission after it has been rebuilt.

There are many different methods used to clean automatic transmission parts. Some rebuilding shops use a parts washing machine, which takes a little time to thoroughly clean a transmission case and associated parts. These parts washers use hot water and a special detergent that is sprayed onto the parts as they rotate inside the cleaner. After cleaning, transmission and parts should be rinsed with water and then air dried before reassembly.

Before removing the torque converter, a measurement may be taken to see how far the torque converter sets back in the bell housing. This measurement can be used during assembly to be sure that the torque converter is fully engaged into the pump. After the case is clean, remove the torque converter and carefully inspect it for damage. Check the converter hub for grooves caused by hardened seals. Also check the bushing contact area. To remove the converter, slowly rotate it as you pull it from the transmission; have a drain pan handy to catch the fluid. It should come out without binding. This is a good time to check the input shaft splines, stator support splines, and the converter's pump drive hub for any wear or damage. Converters with direct drive shafts should be checked to be sure that no excessive play is present at the drive splines of the shaft or the converter. If any play is found in the converter, the converter or the shaft must be replaced.

Position the transmission to perform end-play checks. The transmission end-play checks can provide the technician with information about the condition of internal bearings and seals, as well as clues to possible causes of improper operation found during the road test. These measurements will also determine the thickness of the thrust washers needed during reassembly.

The thrust washers' thickness sets the end-play of various components. Excessive end-play allows clutch drums to move back and forth too much, causing the transmission case to wear. Assembled end-play measurements should be between minimum and maximum specifications, but preferably at the low end of the specifications.

Check all threads in the case for any defects. It may be wise to run a thread chaser through all the threaded holes to clean them and check the condition of the threads. Any threads that are damaged should be repaired, either with a tap or by installing a threaded insert. If a threaded insert cannot be installed and the threads cannot be repaired, the case will have to be replaced.

2. Inspect, measure, and replace oil pump components.

Inspect the pump bore for scoring on both its bottom and sides. A converter that has had a tight fit at the pilot hub could hold the converter drive hub and inner gear too far into the pump, causing cover scoring. A front pump bushing that has too much clearance may allow the gears to run off-center, causing them to wear into the crescent and/or the sides of the pump body. This kind of pump wear investigation should always be followed up by a measurement of a converter hub-to-bushing clearance check. Some transmissions are fitted with *finish-in-place pump bushings*, which means the final bore centerline of the bushing places the inner pump gear in the best location in the pump pocket for pressure development. This means the original factory bronze alloy bushing inside diameter and pump pocket can be offset from its outside diameter. If a remanufactured torque converter with a slightly different hub diameter (smaller) and steel alloy (softer) content are installed, problems can result from rapid wear of the converter hub due to increased clearance that allows the pump inner gear to wander out of its original position, creating noise or low output. If the OE bushing is replaced as a standard procedure, problems can now stem from relocating the inner pump gear centerline from the original pump pocket bore centerline.

The stator support should be inspected for looseness in the pump cover. This can be done while you check for interference inside a torque converter. The shaft's splines and bushings should also be carefully looked at. If the splines are distorted, the shaft and the pump cover should be replaced. The bushings mounted in the hollow stator shaft supporting the turbine shaft and/or the bushings in the hollow turbine shaft supporting the internal oil pump drive shaft control oil flow through the converter and cooler, and their fit must be checked. Bushings must be tight inside hollow shafts and provide the input shaft with a specified clearance.

Inspect the gears and pump parts for deep nicks, burrs, or scratches. Examine the pump housing for abnormal wear patterns. The fit of each gear into the pump body, as well as the centering effect of the front bushing, controls oil pressure loss from the high-pressure side of the pump to the low-pressure input side. Scoring or body wear will greatly reduce this sealing capability. Excessive clearance at any of these mating surfaces mentioned above, including inner-pump gear/rotor-to-hub clearance, can cause or contribute to a condition known as converter drain down.

Torque converter drain down occurs when a vehicle is allowed to sit for a time long enough to allow the fluid to drain back into the reservoir, leaving a low level of fluid in the converter. This void needs to be filled with fluid and the air expelled before the turbine can transmit power to the input shaft of the transmission. The time required to do this causes a delay in vehicle movement after selecting a gear, especially on those transmissions that do not allow filling the torque converter in park. This delay is often expressed as a concern by the customer. Careful examination of these areas will correct or reduce converter drain down and prevent it from becoming a customer concern.

Another important clearance to prevent the possibility of pump cavitation, also called loss of prime by some manufacturers, is pump pocket or gear face clearance. This clearance, if excessive (general rule is 0.002 inch maximum), allows air to be drawn at the neutral (zero pressure) position where the inlet porting in the pump pocket begins. When this condition occurs, the transmission neutralizes, causing momentary or complete and unexplained loss of transmission of power from the torque converter. This normally occurs when fluid temperatures are elevated, which causes reduced viscosity. That, coupled with excessive clearance, allows ambient air to be drawn in, creating the condition.

On fixed displacement pumps, use a feeler gauge to measure the clearance between the outer gear and the pump pocket in the pump housing. Also, check the clearance between the outer pump gear teeth and the crescent, and between the inner gear teeth and the crescent. It is equally important, if specifications are given, to accurately check the inner gear-to-crescent clearance. The pump should be placed on the converter hub with its bushing-to-hub and inner gear-to-drive hub clearances within specifications. This will position the inner gear close to the position that it will run when in operation and allows the possibility of inner gear-to-crescent interference to be viewed. Compare these measurements to the specifications. Use a straightedge and feeler gauge to check gear-side clearance and compare the clearance to the specifications. If the clearance is excessive, replace the pump.

Variable displacement vane-type pumps require different measuring procedures. The inner pump rotor-to-converter drive hub fit, however, is checked in the same way as described for the other pumps. The pump rotor, vanes, and slides are originally selected for size during assembly at the factory. Changing the original size of any of these parts during overhaul can change the sizing and possibly the body of the pump. These parts are available in select sizes for just this reason.

The vanes are subject to wear, as well as cracking and subsequent breakage. The outer edge of the vanes should be rounded, with no flattening. These pumps have an aluminum body and cover halves; any scoring indicates that they should be replaced.

Inspect the reaction shaft's seal rings. If the rings are made of cast iron, check them for nicks, burrs, or uneven patterns, and replace them if they are damaged. Make sure the rings are able to rotate in their grooves. Check the clearance between the reaction shaft support ring groove and the seal ring. If the seal rings are the Teflon™ full circle type, cut them out and use the required tools to replace them.

The outer area of most pumps utilizes a rubber seal. Check the fit of the new seal by making sure the seal sticks out a bit from the groove in the pump. If it does not, it will leak. The seal at the front of the pump is always replaced during overhaul. Most of these seals are the metal-clad lip seal type. Care must be taken to avoid damage to the seating area when removing the old seal.

Check the area behind the seal to be sure the drainback hole is open to the sump. If this hole is restricted, the new seal will possibly blow out. The drainback hole relieves pressure behind the seal. A loose-fitting converter drive hub bushing can also cause the front pump seal to blow out. Finally, pump-mating surfaces must be properly prepped and aligned before final assembly. Surface patterns near bolt holes or passages or witness marks on gaskets for pumps so equipped give indication of cross leaks that are occurring between pump halves. A honing stone can be used to take out minor dents with raised areas. Raised threads on bolt holes retaining pump halves together should be countersunk to prevent half separation during torquing.

Inner and outer gears often have alignment marks that indicate gear orientation, alignment, or rotational direction. Some inner pump gears have their drive tangs offset to indicate orientation. Some outer pump gears are oriented by the size of the radius on their outside diameter. Always study the gears before removal and if necessary, mark them with a common mark across their faces with nonremovable marker or machinist dye to ensure proper reassembly. Pump halves alignment is performed in several ways. Pumps whose covers (stator/reaction shaft support half) are the same diameter as the pump pocket body can be aligned using a large hose clamp or series of clamps or similar special tool, then using dowels to align bolt holes. Some pumps require special tools to align the pump cover, bell housing, and pump body. In addition, some pumps are equipped with alignment pins to give correct alignment. Pumps whose covers (stator/reaction shaft support) are smaller in diameter than the body are typically recessed, which aligns the two halves when bolted together.

3. Check bearing preload; determine needed service.

While a transmission is in operation, the gears, shafts, and bearings are subjected to heat, power transfer loads, and vibrations. Because of this, the drive train must normally be adjusted for the proper fit between parts. These adjustments require the use of precision measuring tools. There are three basic adjustments that are made when reassembling a unit or when a problem suggests that readjustment is necessary. Adjusting the clearance or play between two gears in mesh is referred to as adjusting the backlash. End-play adjustments limit the amount of end-to-end movement of a shaft. Preload is an adjustment made to put a load on an assembly at ambient (room) temperature during assembly to compensate for the heat expansion of shafts, housings, and bearings while at operating temperature.

Backlash is the clearance between two gears in mesh. Excessive backlash can be caused by worn gear teeth, the improper meshing of teeth, or bearings that do not support the gears properly. Excessive backlash can result in severe impact on the gear teeth from sudden stops or directional changes of the gears, which can cause broken gear teeth and gears.

Insufficient backlash causes excessive overload wear on the gear teeth and could cause premature gear failure. Backlash is measured with a dial indicator mounted so that its stem is in line with the rotation of the gear and perpendicular to the angle of the teeth. One gear is then moved in both directions while the other gear is held. The amount of movement on the dial indicator equals the amount of backlash present. The proper placement of shims on a gear shaft is the normal procedure for making backlash adjustments.

End-play refers to the measurable axial or end-to-end looseness of a bearing. End-play is always measured in an unloaded condition. To check end-play, a dial indicator is mounted against the outer side gear or the end of a shaft. The gear or shaft is then pried in one direction all the way and the dial on the indicator zeroed. The gear or shaft is then pried in the opposite direction until movement is stopped and the total movement noted. The amount of movement in thousandths of an inch or tenths of a millimeter is then recorded. That is called total end-play. Shims or adjusting nuts are used to adjust end-play.

When normal operating loads are great, gear trains are often preloaded to reduce the deflection of parts. The amount of preload is specified in service manuals and must be corrected for the design of the bearings and the strength of the parts. If bearings are excessively preloaded, they will heat up and fail. When bearings are set too loose, the shaft will wear rapidly due to the great amounts of deflection it will experience. Gear trains are preloaded by shims, thrust washers, or adjusting nuts, or by using double race bearings. Preload adjustments are normally checked by measuring turning effort with a torque wrench or a pull spring scale attached with a string wound around the component whose turning torque is to be measured. The measurement of turning effort to keep the component moving, rather than the effort to start the component turning, is then recorded and compared to a specification.

4. Check end-play; inspect, measure, and replace thrust washers and bearings as needed.

Before disassembly, a check of end-play becomes important. Not only does that provide a reference measurement, but it can be used to determine the total wear of all thrust washers before disassembly.

The purpose of a thrust washer is to support a thrust load or axial movement and keep parts from rubbing together, thus preventing premature wear on parts, such as planetary gearsets. Selective thrust washers come in various thicknesses to take up clearances and adjust shaft end-play.

Flat washers and bearings should be inspected for scoring, flaking, and wear-through to the base material. Flat washers should also be checked for broken or worn tabs. These tabs are critical for holding the washer in place. On metal flat thrust washers, the tabs may appear cracked at the bend of the tab. This is a normal appearance due to the characteristics of the material used to manufacture them. Plastic thrust washers will not show wear unless they are damaged. The only way to check their wear is to measure the thickness with a micrometer and compare it to a new part. If the plastic thrust washer has any metal embedded into it, the thrust washer should be replaced. All damaged and worn thrust washers and bearings should be replaced.

Use a petroleum jelly-type lubricant to hold thrust washers in place during assembly. This will keep them from falling out of place, which will affect end-play. Besides petroleum jelly, there are special greases designed just for automatic transmission assembly that will also work.

All bearings should be checked for roughness after cleaning. Carefully examine the inner and outer races, and the rollers, needles, or balls for cracks, pitting, etching, or signs of overheating; caged or Torrington needles for bluing, flat spots, or pitting. Needle bluing indicates lack of lubrication or overloading. Cage scuffing (shiny areas between rollers) indicates needle wear requiring replacement.

Tapered roller and ball bearings require visual inspection that will detect obvious damage such as cracks or metal fretting (smearing or relocation of metal creating divots in the surface). Slowly rotating the inner and outer respective races against each other, feeling for varying roughness, will indicate possible noise at operating speed. At times, prying a roller out of the cage of the inner race for visual observation of the inner race will reveal damage that can be the cause of a noise problem. Ball bearings placed in a vise, then squeezed with a slight amount of pressure, may reveal a greater than normal roughness at one or more places in a revolution, which may indicate a faulty race(s).

5. Inspect and replace shafts.

Carefully examine the areas on all shafts that ride in a bushing, bearing, or seal. Also inspect the splines for wear, cracks, or other damage. A quick way to determine spline wear is to fit the mating splines and check for lateral or rotational movement.

Shafts are checked for scoring in the areas where they ride in bushings. Since the shaft is much harder than the bushings, any scoring on the shaft indicates a lack of lubrication at that point. The affected bushing should appear worn into the backing metal. Because shaft-to-bearing fit is critical to correct oil travel throughout the transmission, a scored shaft should be replaced. Lubricating oil is carried through most shafts, and internal inspection for debris is necessary. A blocked oil delivery hole can starve a bushing, resulting in a scored shaft. The internal oil passage of a shaft may not be able to be visually inspected, and only observation during cleaning will give an indication of the openness of the passage. Washing the shaft passage out with a solvent and possibly running a piece of small diameter wire through the passage will dislodge most particles. Be sure to check that the ball closes off the end of the shaft, if the shaft is so equipped, and is securely in place. A missing ball could be the cause of burned planetary gears and scored shafts due to a loss of oil pressure. Any shaft that has an internal bushing should be inspected, as described earlier. Replace all defective parts as necessary.

Input and output shafts can be solid, drilled, or tubular. The solid and drilled shafts are supported by bushings, so the bushing journals of the shaft should be free of noticeable wear at these points. Small scratches can be removed with 300-grit emery paper. Grooved or scored shafts require replacement. The splines should not show any sign of waviness along their length. Check drilled shafts to be sure the drilled portion is free of any foreign material. Wash out the shaft with solvent and run a small-diameter wire through the shaft to dislodge any particles. After running the wire through the opening, wash out the shaft once more and blow it out with compressed air.

If the shaft has a check ball, be certain the ball seats in the correct direction. Some shafts have a ball pressed into one end to block off one end of the shaft. This is used to hold oil in the shaft so the oil is diverted through holes in the side of the shaft. These holes supply oil to bushings, one-way clutches, and planetary gear. If the ball does not fully block the end of the shaft, oil pressure can be lost, causing failure of these components. Some shafts may be used to support another shaft. The output shaft uses the rear of the input shaft to center and support itself. The small bushing found in the front end of the output shaft should always be replaced on these transmissions during rebuilding.

6. Inspect oil delivery circuit, including seal rings, ring grooves, sealing surface areas, feed pipes, orifices, and encapsulated check valves (balls).

At times, leaks may be from sources other than seals. Leakage could be from a worn gasket, loose bolts, cracked housing, or loose line connections. Inspect the outside sealing area of the seal to see if it is wet or dry. If it is wet, see whether the oil is running out or if it is merely a lubricating film. Check both the inner and outer parts of the seals for wet oil.

While removing a seal, inspect the sealing surface, or lips, before cleaning it. Look for the obvious; often overlooked is that the seal is just worn out. Instead of a sharp lip, the point of contact is flattened out due to wear. When compared to a new seal, it is obvious to see, especially if the seal is equipped with a helix to aid in the sealing process. The helix (molded cross hatching) on the seal lip is either bi- or unidirectional. Comparing to a new seal will quickly show the worn area and the level of wear. Warpage, cuts, and gouges, or particles embedded in the seal are all indications of a faulty seal. On spring-loaded lip seals, make sure that the spring is seated around the lip and that the lip was not damaged when first installed. If the seal's lip is hardened, this was probably caused by heat from either the shaft or the fluid.

If the seal is damaged, check all shafts for roughness, especially at the seal contact area. Look for deep scratches or nicks that could have damaged the seal. Determine if the shaft spline, keyway, or a burred end could have caused a nick or cut in the seal lip during installation. Inspect the bore into which the seal was fitted. Look for nicks and gouges that could create a path of oil leakage. A coarsely machined bore can allow oil to seep out through a spiral path. Sharp corners at the bore edges can score the metal case of the seal when it is installed. These scores can make a path for oil leakage. All check balls should be cleaned and dried with compressed air before reassembly.

7. Inspect and replace bushings.

Bushings should be inspected for pitting and scoring. Always check the depth to which bushings are installed and the direction of the oil groove, if so equipped, before you remove them. Many bushings that are used in the planetary gearing and output shaft area have oiling holes in them. Be sure to line these up correctly during installation to prevent blocking off oil delivery holes that could destroy the gear train. If any damage is evident on the bushing, it should be replaced.

Bushing wear can be checked directly, as well as checked by observing the lateral movement of the shaft that fits into the bushing. Any noticeable lateral movement indicates wear, and the bushing should be replaced. The amount of clearance between the shaft and the bushing can be checked with a wire-type feeler gauge. Insert the wire between the shaft and the bushing. If the gap is greater than the maximum allowable gap, the bushing should be replaced. Measuring the inside diameter of the bushing and the outside diameter of the shaft with a vernier caliper or micrometer is a way to check this fit. This is a critical fit throughout the transmission and especially at the converter drive hub.

Most bushings are press fit into a bore. To remove them, they are driven out of the bore with a properly sized bushing tool. Some bushings can be removed with a slide hammer fitted with an expanded or threaded fixture that grips to the inside of the bushing and collapses it. Once collapsed, the bushing can easily be removed with a pair of pliers. Small-bore bushings that are located in areas where it is difficult to use a bushing tool can be removed by tapping the inside bore of the bushing with threads that match a selected bolt, which fits into the bushing. After the bushing has been tapped, insert the bolt and use a slide hammer to pull the bolt and bushing out of its bore.

Whenever possible, all new bushings should be installed with the proper bushing driver. An arbor press is the tool of choice for bushing installation, especially if the bushing has a large amount of press (interference) fit into its bore.

If an arbor press is not available, and the bushing must be driven in, it is possible that the bushing will swell at the driven end due to the pressure of the driver and the force of the hammer blows. If so, a bearing knife can be used to chamfer the edge of the bushing to remove any material raised by installation. The use of these tools prevents damage to the bushing and allows for proper seating of the bushing into its bore.

8. Inspect and measure planetary gear assembly; replace parts as necessary.

A close inspection of the planetary gearset is a must to eliminate the possibility of causing noises in a newly rebuilt unit. All planetary gear teeth should be inspected for chips or stripped teeth. Any gear that is mounted to a splined shaft must have its splines checked for mutilation or shifted splines. The planetary gears used in automatic transmissions are helical-type gears, like the ones used in most manual transmissions. This type of gear provides low noise in operation, but makes it necessary to check the end-play of individual gears during inspection. The helical cut makes the gears thrust to one side during inspection. This can put a lot of load on the thrust washers and may wear them beyond specification.

Look first for obvious problems like blackened gears or pinion shafts. These conditions indicate severe overloading and require that the carrier be replaced. Occasionally, the pinion gear and shaft assembly can be replaced individually. When looking at the gears themselves, note that a bluish color can be a normal condition, as this is part of a heat-treating process used during manufacturing. With the gearset assembled, inspect the mating of the gears. Excessive backlash between gears indicates excessive wear and the unit should be replaced. The gear carrier should have no cracks or other problems. Replace any abnormal or worn parts. Check each gear individually by rolling it on its shaft to feel for looseness, roughness, or binding of the needle bearings. Wiggle the gear to be sure it is not loose on the shaft and to feel for roughness or binding of the needle bearings. Looseness will cause the gear to whine when it is loaded. The end-play of the pinion gear should be checked by inserting a feeler gauge between the gear and carrier. Some gearsets can be disassembled, and shims installed to correct improper end-play. On some Ravigneaux units, the clearance at both ends of the long pinion gears must also be checked and compared to specifications.

As a general rule, gear train end-play gives quietest operation when its end-play is kept at the low end of the specification range. Also, inspect the gears and teeth for chips or imperfections, as these will also cause whine. Check any thrust surfaces for damage and replace if necessary. Some planetary carrier assemblies may contain a Torrington bearing just below the planetary gears that cannot be removed. Inspect this bearing for damage, and if faulty, replace the assembly. Check the gear teeth around the inside of the planetary ring gear. Examine thrust washers and thrust surfaces for any damage that may be cause for replacement.

Sun gears should be inspected similarly to the previous components as far as chipped teeth, cracks, discoloration, and thrust surfaces. Many sun gears have inner bushings that should be inspected for looseness on their respective shafts.

Any of the previously mentioned components—sun gear, planetary carrier assembly, and the planetary ring gear—may have other components attached to their housings. These components can include splines for clutch discs or shafts. Drums or hubs can also be part of the gearset. These components should be inspected for wear. Clutch discs should slide easily on their respective splines. Drums/shells, hubs, or shafts that are splined to the gearset should fit tightly and not move when installed. Check drums/shells that commonly fit to the sun gear closely, as they will typically be damaged in this area.

9. Inspect case bores, passages, bushings, vents, mating surfaces, thread condition, and dowel pins; repair or replace as necessary.

The transmission case should be thoroughly cleaned and all passages blown out. After the case has been cleaned, all the bushings, fluid passages, bolt threads, clutch plate splines, and the governor bore should be checked. The passages can be checked for restrictions and leaks by applying compressed air to each one. If the air comes out the outer end, there is no restriction. To check for leaks, plug off one end of the passage and apply air to the other. If pressure builds in that passage, there are probably no leaks in it.

Modern transmission cases are made of aluminum, primarily to save weight. Aluminum is a soft material that can be deformed, scratched, cracked, or scored much more easily than cast iron. Special attention should be given to the following areas: the clutch, the oil pump, the servo, and the accumulator bores. All bores should be smooth to avoid scratching or tearing the seals. The servo piston could also hang up in a bore that is deeply scored, which could cause a shifting concern for the customer. Check the fit of the servo piston in the bore without the seal, if possible, to be sure it has free travel.

There should be no tight spots or binding over the whole range of travel. Any deep scratches or gouges that cause binding of the piston, which causes line pressure loss that can set DTCs on EATs and/or shifting concerns. In the past, when a case had scoring of this sort in servo and accumulator bores, it would have to be scrapped. However, repair sleeves currently are available from manufacturers to prevent expensive case replacement.

Case-mounted accumulator bores are checked the same as servo bores. The oil pump bore at the front of the case should be free of any scratches that would keep the o-ring from sealing the outer diameter of the pump to the front of the case. Case-mounted hydraulic clutch bores are prone to the same problems as servo bores. Look for any scratches or gouges in the sealing area that would affect the rubber seals. It is possible to damage these areas during disassembly, so be careful with tools used during overhaul.

Sealing surfaces of the case should be inspected for surface roughness, nicks, or scratches where the seals ride. Any problems found in servo bores, clutch drum bores, or governor support bores can cause pressure leakage in the affected circuit.

Imperfections in steel or cast-iron parts can usually be polished out with crocus cloth. Care should be taken so as not to disturb the original shape of the bore. Under no circumstances should sandpaper be used. Sandpaper will leave too deep a scratch in the surface. Use a crocus cloth inside clutch drums to remove the polished marks left by the cast-iron sealing rings. This will help the new rings rotate with the drum as designed. As a rule, all sealing rings, either cast iron or Teflon™, are replaced during overhaul, as this gives the desired sealing surface required for proper operation.

Passages in the case guide the flow of fluid through the case. Although not common, porosity in this area can cause cross-tracking of one circuit to another. This can cause bind-up (two gears at once) or a slow bleed of pressure in the affected circuit, which can lead to slow burnout of a clutch or band. If this is suspected, try filling the circuit with solvent and watching to see if the solvent disappears or leaks away. If the solvent goes down, you will have to check each part of the circuit to find where the leak is. Be sure to check that all necessary check balls were in position during disassembly.

Small screens found during teardown should be inspected for foreign material. These screens are used to prevent valve hang-up at the pressure regulator and governor. Most screens can be removed easily. Care should be taken when cleaning because some solvents will destroy the plastic screens. Low air pressure can be used to blow the screens out in a reverse direction.

Bushings in a transmission case are normally found in the rear of the case and require the same inspection and replacement techniques as other bushings in the transmission. Always be sure that the oil passages to a pressure-fed bushing or bearing are open and free of dirt and foreign material. It does no good to replace a bushing without checking to be sure it has good oil flow.

Vents are located in the pump body or transmission case and provide for equalization of pressure in the transmission. The symptoms of a restricted transmission vent are that of a weak pump with low volume output. A simple check that can be performed is to flow check the cooler circuit while placing a slight amount of pressure through the fill tube with an air nozzle.

If flow increases significantly a restricted transmission vent is suspect. These vents can be checked by blowing low-pressure air through them, squirting solvent brake cleaning spray through them, or by pushing a small diameter wire through the vent passage. A clean, open passage is all you need to verify proper operation.

10. Inspect and replace transaxle drive chains, sprockets, gears, bearings, and bushings.

The drive chains used in some transaxles should be inspected for side-play and stretch. These checks are made during disassembly and should be repeated as a double-check during reassembly. On some transaxles, chain deflection is measured between the centers of the two sprockets. Typically, very little deflection is allowed.

One manufacturer has the technician deflect the chain inward on one side until it is tight. Mark the housing at the point of maximum deflection. Then deflect the chain outward on the same side until it is tight. Again mark the housing in line with the outer edge of the chain at the point of maximum deflection. Measure the distance between the two marks. If this distance exceeds specifications, replace the drive chain. Other manufacturers' procedures vary.

Be sure to check for an identification mark on the chain during disassembly. These can be painted or have dark-colored links, which may indicate either the top or the bottom of the chain. The chain should be reinstalled in the exact same position as it was disassembled.

The sprockets should be inspected for tooth wear at the point where they ride. If the chain was found to be slack, it may have worn the sprockets in the same manner as the engine timing gears wear when the timing chain stretches. A slightly polished appearance on the face of the gears is normal.

The bearings and bushings used on the sprockets need to be checked for damage. The radial needle thrust bearings must be checked for any deterioration of the needles and cage. The running surface in the sprocket must also be checked, as the needles may pound into the gear's surface during operation. The bushings should be checked for any signs of scoring, flaking, or wear. Replace any defective parts. The removal and installation of the chain drive assembly of some transaxles requires that the sprockets be spread slightly apart. The key to doing this is to spread the sprockets just the correct amount. If they are spread too far, they will not be easy to install or remove.

11. Inspect, measure, repair, adjust, or replace transaxle final drive components.

Transaxle final drive units should be carefully inspected. Examine each gear, thrust washer, and shaft for signs of damage. If the gears are chipped or broken, they should be replaced. Also inspect the gears for signs of overheating or scoring on the bearing surface of the gears.

Final drive units may be hypoid gear, helical gear, or planetary gear units. The hypoid and helical gear type should be checked for worn or chipped teeth, overloaded tapered roller bearings, and excessive differential side gear and differential pinion gear wear. Excessive play in the differential is a cause of engagement clunk. Be sure to measure the clearance between the side gears and the differential case, and to check the fit of the differential pinion gears on their shaft. Proper clearances can be found in the appropriate shop manual. It is possible that the side bearings of some final drive units are preloaded with shims. Select the correct size shim to bring the unit into specifications. With a torque wrench, measure the amount of rotating torque. Compare your readings against specifications.

If the ring gear carrier bearing preload and case gears end-play are within spec and the bearings are in good condition, the parts can be reused. However, always install new seals during assembly. It should be noted that these tapered roller bearings function the same as RWD rear-axle side bearings and should not set the preload to the specifications for a new bearing. Used bearings should be set to the amount found during teardown or about one-half the preload of a new bearing.

Planetary-type final drives are also checked for the same differential case problems that the helical type would encounter. The planetary pinion gears need to be checked for looseness or roughness on their shafts and for end-play. Any problems found normally result in the replacement of the carrier as a unit since most pinion bearings and shafts are sold as separate parts. Again, specifications for these parts are found in the shop manual.

Planetary-type final drives, like helical final drives, are available in more than one possible ratio for a given type of transaxle, so care should be taken to assure that the same gear ratios are used during assembly. This is not normally a problem when overhauling a single unit; however, in a shop where many transmissions are being repaired, it is possible to mix up parts, causing problems during the rebuild.

12. Assemble after repair.

Once the transmission has been cleaned, inspected, and all faulty components repaired, assembly can begin. During assembly, it is important to keep all components as clean as possible. Dirt or debris could clog fluid passages and cause valves to stick. Soak all bands and clutch discs in ATF for 30 minutes. Prelube all o-rings and seals with an approved assembly lubricant; never install a dry seal. Lube all internal components that will have metal-to-metal contact such as the bearings and gearsets. Overuse of sealers or gasket maker should be avoided if used at all. The excess can break off inside the transmission and clog internal passages. Be sure to refer to service information and make any necessary clearance or end-play checks. These clearances are often adjusted with shims or thrust washers. Also be sure that clutch pack clearances are within spec and the bands are properly adjusted. Clutch pack clearances are typically adjusted by installing different thickness snap rings or reaction plates. Many bands are adjusted with an adjustment screw or a selective servo pin.

When installing the pump, it may be helpful to use guide pins and seal compressors to ease installation. Before installing the valve body, it is a good idea to perform an air test on the clutches and bands. This involves applying low-pressure shop air to the hydraulic passages of the bands and clutches. If they are in good working order, a clunk will be heard as the unit engages. A hissing noise indicates a leak in the circuit. When installing the valve body, hold all check balls in place with assembly lube and torque the valve body bolts in the proper sequence. With the transmission assembled, the torque converter can be installed. It is common practice to pour approximately one quart of transmission fluid into the torque converter before installation. During installation, be sure that the torque converter completely engages the pump. A measurement of how deep the torque converter sets into the bell housing should have been taken before removal. This measurement can be used upon installation to be sure that the torque converter is fully engaged. Failure to fully engage the torque converter can result in a damaged torque converter drive hub and/or pump.

3. Friction and Reaction Units (4 Questions)

▨ 1. Inspect hydraulic clutch pack assembly; replace parts as necessary.

The hydraulic clutch pack assembly consists of component parts such as steel plates, friction plates, hydraulic piston and bore, springs, hubs, etc. Once a clutch assembly has been taken apart, you may wish to inspect the clutch components or continue to disassemble the remainder of the clutch units in the transmission. If you choose the latter, make sure you keep the parts of each clutch separate from the others.

Clean the components of the clutch assembly. Make sure all clutch parts are free of any residue of varnish, burned disc facing material, or steel filings. Take special care to wash out any foreign material from the inside of the drums and hub disc splines. If left in, the material can be washed out by the fresh transmission fluid and sent through the transmission. This can ruin the rebuild.

The clutch's hub splines must be in good shape with no excessively rounded corners or shifted splines. Test their fit by trial-fitting one new clutch disc on the splines. Move the discs up and down the splines to check for binding. If they bind, this can cause dragging of the discs during a time when they should be free-floating.

Replace the component whose hub caused the disc to drag during this check. Check the spring retainer; it should be free-floating. Check the spring retainer; it should be flat and not distorted at its inner circumference. Check all springs for height, cracks, and straightness. Any springs that are not the correct height or that are distorted should be replaced. Many retainers have springs attached to them by crimping. This speeds up production at the assembly line. Turning this type of retainer upside down is a quick check of spring length. Closely examine the Belleville spring for signs of overheating or cracking, and replace it if it is damaged.

The steel plates should be checked to be sure they are flat and not worn too thin. Check all steel plates against the thickest one in the pack or a new one. A visual inspection starts with checking if spot scuffing or bluing is present on the plate; if it is, discard it. However, some discoloration of the plate is okay, and the plate is suitable for reuse as long as the plate is not warped. Warping can be checked by stacking plates one on top of another and checking for gaps around the outside diameter. Be aware that some transmission's steel plates are slightly dished and others can be a wave design when checked on the inside diameter.

The dish or wave shape acts as a cushion to soften the clutch application feel. Next, check for scoring and damaged drive lugs; if present, discard the plate. Most steel plates will have an identification notch or mark on the outer tabs to inform the transmission manufacturer who the supplier is of the plates. If the plates pass inspection, remove the polished surface finish so the steel plates are ready for reuse. The steel plates should also be checked for flatness by placing one plate on top of the other and checking the shape on the inside and outside diameters. Clutch plates must not be warped or cone shaped. Also, check the steel plates for burning and scoring, and for damaged driving lugs. Check the grooves inside the clutch drum and check the fit of the steel plates, which should travel freely in the grooves.

Close inspection of the friction discs is simple. The disc will show the same type of wear as bands will. Disc facing should be free of chunking, flaking, and burned or blackened surfaces. Discs that are stripped of their facing have been overheated and subject to abuse.

In some cases, the friction discs and steel plates can be welded together. This occurs when extreme heat melts the facing's bonding and the facing separates from the disc.

As the facing comes off, metal-to-metal contact is made and the disc and plate fuse together. This may lock the clutch in an engaged condition. Depending on which clutch is affected, drivability problems can include driving in neutral, binding up in reverse, starting in direct drive, binding up in second, and other problems that are not that common.

If the discs do not show any signs of deterioration, squeeze each disc to see if fluid is still trapped in the facing material. If fluid comes to the surface, the disc is not glazed. Glazing seals off the surface of the disc and prevents it from holding fluid. Holding fluid is basic to proper disc operation. It allows the disc to survive engagement heat, which would otherwise burn the facing and cause glazing. Fluid stored in the friction material cools and lubricates the facing as it transfers heat to the steel plate and also carries heat away as some oil is spun out of the clutch pack by centrifugal force. This helps avoid the scorching and burning of the disc. The clutch disc must not be charred, glazed, or heavily pitted.

If a disc shows signs of flaking or if friction material can be scraped off easily, replace the disc. A black line around the center of the friction surface also indicates that the disc should be replaced. Examine the teeth on the inside of each friction disc for wear and other damage.

Wave plates are used in some clutch assemblies to cushion the application of the clutch. These should be inspected for cracks and other damage. Never mix wave plates from one clutch assembly with another. As an aid in assembly, most wave plates will have different identifying marks.

2. Measure and adjust clutch pack clearance.

The clearance check of a clutch pack is critical for correct transmission operation. Excessive clearance causes delayed gear engagements, while too little clearance causes the clutch to drag. Adjusting the clearance of multiple-disc clutches can be done with the large outer snap ring in place.

With the clutch pack and pressure plate installed, use a feeler gauge to check the distance between the pressure plate and the outer snap ring. Clearances can also be measured between the backing plate and the uppermost friction disc. If the clutch pack has a waved snap ring, place the feeler gauge between the flat pressure plate and the wave of the snap ring farthest away from the pressure plate. Compare the distance to specifications. Attempt to set the pack clearance to the smallest dimension shown in the chart.

Clearance can also be checked with a dial indicator and hook tool. The hook tool is used to raise one disc from its downward position, and the amount that it is able to move is recorded on the dial indicator. This represents the clearance.

Multiple-disc clutch packs have several methods of clearance adjustment. Selective (varying thicknesses) components are used to adjust the free movement of the clutch pack. These range from retaining plate snap rings and pressure and reaction plates to custom steel plate thicknesses usually offered in overhaul kits. If clearance is less than specified, thicker selective components are used to take up the clearance; the opposite is true if the clearance is too great. Some transmissions use tapered snap rings for additional retaining power against the hit of the reaction or pressure plate when the clutch is applied. Snap-ring orientation and end location are critical to the ring's ability to hold and also to its longevity.

3. Air test the operation of clutch and servo assemblies.

After the clearance of the clutch pack is set, perform an air test on each clutch. This test will verify that all of the seals and check balls in the hydraulic components are able to hold and release pressure.

Air checks can also be made with the transmission assembled. This is the absolute best way to check the condition of the circuit because there are very few components missing from the circuit. The manufacturers of different transmissions have designed test plates that are available to test different hydraulic circuits. Testing with the transmission assembled also allows for testing the servos.

The clutch assemblies in many transmissions can be air tested by installing the oil pump assembly with its reaction shaft support over the input shaft and sliding it into place on the front clutch drum. When the clutch drums are mounted on the oil pump, all components in the circuit can be checked. If the clutch cannot be checked in this manner, blocking off apply ports with your finger and applying air pressure through the other clutch apply port will work.

Invert the entire assembly and place it in an open vise or transmission support tool. Then air test the circuit using the test hole designated for that clutch. Be sure to use low-pressure compressed air (25–35 psi) to avoid damage to the seals. High-pressure air may blow the rubber seals out of the bore or roll them on the piston.

While applying air pressure, you may notice some air escaping at the metal of the Teflon™ seal areas. This is normal, as these seals have a controlled amount of leakage designed into them. This is also true with some piston-applied clutch packs. What is most important is that a *thud* is heard, indicating the piston has moved applying the clutch pack. Experience will help you determine what is a normal amount of air by-passing (hissing) at these apply pressures with these clutch assemblies. It should release quickly without any delay or binding. If no evidence of piston movement is heard, examine the check ball seat for evidence of air leakage.

4. Inspect one-way clutch assemblies; replace parts as necessary.

Because they are purely mechanical in nature, one-way clutches are relatively simple to inspect and test. The durability of these clutches relies on constant fluid flow during operation. If a one-way clutch has failed, a thorough inspection of the hydraulic feed circuit to the clutch must be made to determine if the failure was due to fluid starvation. The rollers and sprags ride on an overrunning state, and any loss of fluid can cause a rapid failure of the components. Sprags, by design, produce the fluid wave effect as they slide across the inner and outer races, making them somewhat less prone to damage. Rollers, due to their spinning action, tend to throw off fluid, which allows more chance for damage during fluid starvation. During the check of the hydraulic circuit, take a look at the feed holes in the races of the clutch. Use a small-diameter wire and spray carburetor cleaner or brake cleaner to be certain the feed holes are clear. Push the wire through the feed holes and spray the cleaner into them. Blowing through them with compressed air after cleaning is recommended.

Roller clutches should be disassembled to inspect the individual pieces. The surface of the rollers should have a smooth finish with no evidence of any flatness. Likewise, the race should be smooth and show no sign of brinelling, as this indicates severe impact loading. This condition may also cause the roller clutch to buzz as it overruns.

All rollers and races that show any type of damage or surface irregularities should be replaced. Check the folded springs for cracks, broken ends, or flattening out. All of the springs from a clutch assembly should have approximately the same shape. Replace all distorted or otherwise damaged springs. The cam surface, like the smooth race, must be free of all irregularities.

Sprag clutches cannot easily be disassembled, so a complete and thorough inspection of the assembly is necessary. Pay particular attention to the faces of the sprags. If the faces are damaged, the clutch unit should be replaced. Sprags and races with scored or torn faces are an indication of dry running and require the replacement of the complete unit.

Once the one-way clutch is ready for installation, verify that it overruns in the proper direction. In some cases, it is possible to install the clutch backward, which would cause it to overrun and lock in the wrong direction. One-way clutches are often used to provide additional holding power for clutches and bands, or to allow a seamless transfer of power source between engine and vehicle (preventing planetary freewheeling for engine braking in lower ranges), and to allow freewheeling of planetary members and attached gear train (overdrive range during deceleration) to prevent engine braking, which helps fuel economy. Improper installation (backward) can cause some strange drivability problems; therefore, a good understanding of power flow through the one-way clutch is needed for accurate diagnosis and correct transmission assembly. To make sure you have installed the clutch in the correct direction, determine the direction of lockup before installing the clutch. One-way clutch assemblies are sometimes the source of a noise from the transmission in a particular range or mode of operation (e.g., decelerating). This noise is usually a buzzing noise. Knowing the power flow of the transmission in each range will help in diagnosing the one-way clutch as the noise source as to whether it is in a holding (no movement between its races) or overrunning mode of operation.

5. Inspect bands and drums (housings/cylinders); replace and/or adjust as necessary.

Servicing bands and their components includes inspection of the bands, as well as the drums that the bands wrap around. Before the introduction of overdrive automatic transmissions, most bands operated in free condition during most driving conditions. This means the band was not applied in the cruising gear range. However, many overdrive automatic transmissions use a band in the overdrive cruise range, which puts an additional load on the band and subsequently causes additional wear on the band. For this reason, a thorough inspection of the bands is very important. It must be able to hold or provide braking action.

The bands in a transmission will be either single or double wrap, depending on the application. Both types can be the heavy-duty cast-iron type or the normal strap type. The friction material used on clutches and bands is quite absorbent. This characteristic can be used to tell if there is much life left in the lining. Simply squeeze the lining with your fingers to see if any fluid appears. If fluid appears, this tells you the lining can still hold fluid and has some life left in it. It is hard to tell exactly how long the band will last, but at least you have an indication that it is still useable. Strap- or flex-type bands should never be twisted or flattened out. This may crack the lining and lead to flaking of the lining.

Band failure found during an overhaul is easy to spot. Look for chipping, cracks, burn marks, glazing, and nonuniform wear patterns and flaking. If any of these defects are apparent, the band should be replaced.

Also inspect brake band friction material for wear. If the linings have wear, carefully check the band struts, levers, and anchors for wear. Replace any worn or damaged parts. Look at the linings of heavy-duty bands to see if the lining is worn evenly. A twisted band will show taper wear on the lining. If the friction material is blackened, this is caused by an excessive buildup of heat. High heat may weaken the bonding of the lining and allow the lining to come loose from the metal portion of the band.

The drum surface should be checked for discoloration, scoring, glazing, and distortion. The drums will be either iron castings or steel stampings. Cast-iron drums that are not scored can generally be restored to service by sanding the running surface with 180-grit emery paper in the drum's normal direction of rotation. A polished surface is not desirable on cast-iron drums.

The surface of the drum must also be flat. This is not usually a problem with a cast-iron drum, but it can affect the stamped steel-type drum. It is possible for the outer surface of the drum to dish outward during its normal service life. Check the drum for flatness across the outer surface where the band runs. Any dishing here will cause the band to distort as it attempts to get a full grip on the drum. Distortion of the band weakens the bond of the friction material to the band and will cause early failure due to flaking of the friction lining. A dished stamped steel drum should be replaced. Check the service manual for maximum allowable tolerances.

Sample Preparation Exams

INTRODUCTION

Included in this section are a series of six individual preparation exams that you can use to help determine your overall readiness to attempt and successfully pass the Automatic Transmission/ Transaxle (A2) ASE certification exam. Located in Section 7 of this book you will find blank answer sheet forms that you can use to designate your answers as you attempt each of the preparation exams. Using these blank forms will allow you to attempt each of the six individual exams multiple times without risk of viewing your prior responses.

Upon completion of each preparation exam, you can determine your exam score using the answer keys and explanations located in Section 6 of this book. Included in the explanation section for each question is the specific task area being assessed by that individual question. This additional reference information may prove useful if you need to refer back to the task list located in Section 4 for additional support.

PREPARATION EXAM 1

1. A late-model vehicle needs to be road tested to verify a shifting concern. Technician A says that the customer complaint should be read carefully prior to beginning the road test. Technician B says that the transmission fluid should be checked prior to beginning the road test. Who is correct?

   A. A only
   B. B only
   C. Both A and B
   D. Neither A nor B

2. Which of the following faults would be most likely to cause a shudder condition during the road test?

   A. Torque converter lockup clutch
   B. Weak transmission oil pump
   C. Planetary gearset chipped
   D. Torque converter stator slipping

3. Technician A says that a faulty constant velocity joint can cause a clicking noise when the vehicle is making a turn. Technician B says that a faulty wheel hub bearing can cause a metallic roar while making a turn. Who is correct?

   A. A only

   B. B only

   C. Both A and B

   D. Neither A nor B

4. Which of the following faults would most likely cause a vibration at 50 to 60 mph?

   A. Low line pressure

   B. Slipping clutch pack

   C. Universal joint

   D. Output shaft seal

5. A vehicle is being diagnosed for a transmission fluid leak near the bell housing area. Technician A says that a failed front pump seal could cause a leak in this area. Technician B says that a leaking transmission cooler could cause a leak in this area. Who is correct?

   A. A only

   B. B only

   C. Both A and B

   D. Neither A nor B

6. All of the following problems could cause a low transmission fluid level EXCEPT:

   A. Ruptured transmission cooler

   B. Slipping clutch pack

   C. Faulty modulator valve

   D. Front pump seal

7. Which of the following conditions could cause a vehicle to have reverse and first gears but not any up-shifts?

   A. Low line pressure

   B. Excessive line pressure

   C. Low governor valve pressure

   D. Excessive servo pressure

8. Which gear should a technician put the transmission in to create maximum line pressure?

   A. Manual second

   B. Drive

   C. Park

   D. Reverse

9. A vehicle has a stall speed that is 350 rpm below the standard specification. Technician A says that a slipping clutch pack could be the cause. Technician B says that a plugged exhaust system could be the cause. Who is correct?

    A.   A only
    B.   B only
    C.   Both A and B
    D.   Neither A nor B

10. Technician A says that a stall test should be performed before the engine warms up. Technician B says performing a stall test for a minimum of two consecutive minutes is recommended. Who is correct?

    A.   A only
    B.   B only
    C.   Both A and B
    D.   Neither A nor B

11. Technician A says that the engine RPM should drop 600 rpm when the torque converter clutch engages. Technician B says that the torque converter should disengage when the brake pedal is depressed. Who is correct?

    A.   A only
    B.   B only
    C.   Both A and B
    D.   Neither A nor B

12. A vehicle with an electronic shifted transmission is being road tested. Technician A says that a scan tool can be useful to use during the road test. Technician B says that a pressure manifold can be useful during the road test. Who is correct?

    A.   A only
    B.   B only
    C.   Both A and B
    D.   Neither A nor B

13. Which of the following activities would be LEAST LIKELY to be performed during the road test of a vehicle with an electronic transmission?

    A.   Monitor the shift points of all forward gears.
    B.   Monitor the vacuum level to the modulator.
    C.   Listen for unusual noises in the drive train.
    D.   Observe the feel of the downshift pattern.

14. A pressure control solenoid has failed. This may cause all of the following EXCEPT:

    A. A DTC will set.
    B. The main line pressure will increase.
    C. The transmission will shift hard.
    D. The transmission will have no forward gears.

15. The pressure control solenoid connector was left unhooked following a transmission installation. Technician A says that the transmission will likely jerk when it is shifted from park to reverse. Technician B says that the transmission will likely jerk when shifted from park to drive. Who is correct?

    A. A only
    B. B only
    C. Both A and B
    D. Neither A nor B

16. All of the following conditions have to be met in order for the torque converter clutch to engage EXCEPT:

    A. Brakes are not applied.
    B. Vehicle speed minimum is met.
    C. Transmission is in first gear.
    D. Engine temperature minimum is met.

17. A vehicle is being diagnosed for a drop in fuel economy. Technician A says that an inoperative torque converter clutch could be the cause. Technician B says that a stuck open engine thermostat could be the cause. Who is correct?

    A. A only
    B. B only
    C. Both A and B
    D. Neither A nor B

18. A transmission fluid temperature (TFT) sensor is being tested. Technician A says that an ammeter should be used to test this device. Technician B says that the resistance of the sensor should decrease as the sensor is heated. Who is correct?

    A. A only
    B. B only
    C. Both A and B
    D. Neither A nor B

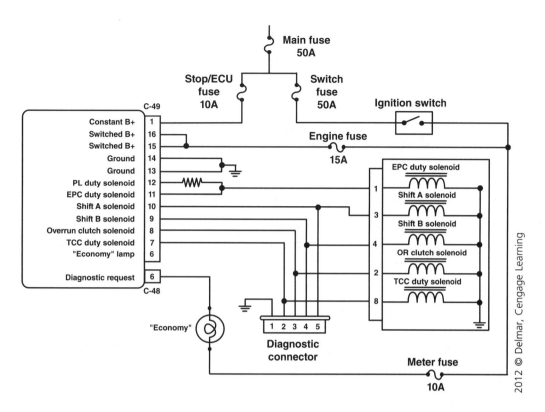

19. Referring to the figure above, Technician A says that all of the solenoids receive ground from the transmission case. Technician B says that the transmission EPC duty solenoid receives power from pins 11 and 12 of the transmission computer. Who is correct?

    A. A only

    B. B only

    C. Both A and B

    D. Neither A nor B

20. The generator needs to be replaced on a late-model vehicle. Technician A says that the replacement generator should have the same diameter pulley. Technician B says that the replacement generator should be disassembled and tested prior to installing on the vehicle. Who is correct?

    A. A only

    B. B only

    C. Both A and B

    D. Neither A nor B

21. Which of the following conditions would be the most likely condition to cause the generator to overcharge?

    A. Blown fusible link in charge circuit

    B. An open rotor coil

    C. A full-fielded rotor

    D. A faulty voltage regulator

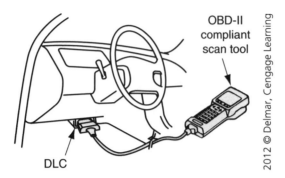

22. The tool shown in the picture above can be used for which of the following purposes?

    A. Checking for the supply voltage (B+) at the transmission computer
    B. Accessing the diagnostic trouble codes (DTCs) in the transmission computer
    C. Testing the resistance of the speeds sensor coil
    D. Testing the voltage of the speeds sensor

23. A vehicle is being diagnosed for a jerking condition that is most prevalent in overdrive. Technician A says that a faulty spark plug could be the cause. Technician B says that a slipping clutch pack in the transmission could be the cause. Who is correct?

    A. A only
    B. B only
    C. Both A and B
    D. Neither A nor B

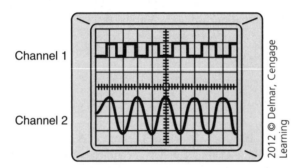

24. Referring to the figure above, Technician A says the pattern on Channel 1 of the oscilloscope is a good signal from a permanent magnet generator (speed sensor). Technician B says the pattern from Channel 2 is a good signal from a "hall effect" sensor. Who is correct?

    A. A only
    B. B only
    C. Both A and B
    D. Neither A nor B

25. Technician A says that a scan tool can be used to retrieve trouble codes from the transmission computer. Technician B says that a scan tool can be used to retrieve sensor data from the transmission computer. Who is correct?

    A. A only
    B. B only
    C. Both A and B
    D. Neither A nor B

26. A vehicle is being diagnosed for a problem of the transmission not engaging into manual low gear. The transmission will engage into all other gears without a problem. Technician A says that the shift indicator could be misaligned. Technician B says that the shift linkage could be binding. Who is correct?

    A. A only
    B. B only
    C. Both A and B
    D. Neither A nor B

27. A transaxle is being diagnosed for a late shifting problem. Which of the following faults would most likely cause this problem?

    A. Broken throttle valve cable
    B. Throttle valve linkage adjusted too tight
    C. Loose band adjustment
    D. Throttle valve linkage adjusted too loose

28. A transmission is leaking near the area where the drive shaft enters the transmission. Which of the components below is the most likely cause of this leak?

    A. Oil pump seal
    B. Output shaft seal
    C. Manual linkage seal
    D. Input speed sensor seal

29. A transmission is being diagnosed for a problem of repeated output shaft seal failures. Which of the following conditions would most likely cause this problem?

    A. Worn differential drive pinion
    B. Overfilled transmission
    C. Worn U-joints at the differential end of the drive shaft
    D. Worn output shaft support bushing

30. The windshield fogs up when the defroster is turned on and the cab is filled with a sweet smell. Which of the following is the most likely cause?

    A. Blown head gasket
    B. Leaking heater control valve
    C. Leaking evaporator core
    D. Leaking heater core

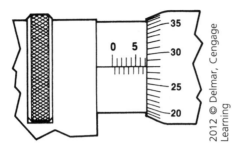

2012 © Delmar, Cengage Learning

31. What is the measurement reading on the metric micrometer above?

    A. 9.28 mm

    B. 7.78 mm

    C. 7.28 mm

    D. 9.78 mm

32. Which of the following practices would be LEAST LIKELY to be followed when installing the bolts into a valve body?

    A. Start all bolts by hand.

    B. Tighten all bolts gradually.

    C. Apply thread locker to all valve body bolt threads.

    D. Use the specified torque wrench to complete the tightening sequence.

33. Technician A says that a servo can be air tested to test for proper piston sealing. Technician B says that the servo spring causes the servo to release when pressure is released from the piston. Who is correct?

    A. A only

    B. B only

    C. Both A and B

    D. Neither A nor B

34. Which of the following details would be the most likely item to be located on a wiring diagram for an electronic transmission?

    A. The location of a component

    B. A flowchart for troubleshooting an electrical problem

    C. The current rating of a circuit

    D. The color and circuit number of a wire

35. Which of the following definitions best describes a connector?

    A. A magnetic switch

    B. A metallic component that is used to tie electrical circuits together

    C. A plastic housing that is used to hold terminals as well as to plug into electrical items

    D. A joint where two or more electrical circuits connect to each other

36. What would most likely cause the engine to twist excessively when accelerating quickly?

    A. Loose transmission bell housing
    B. Broken transmission mount
    C. Loose engine accessory bracket
    D. Broken engine mount

37. Which gear is most likely to be selected when checking the transmission fluid level?

    A. Manual low
    B. Reverse
    C. Neutral
    D. Overdrive

38. Which of the following components would be LEAST LIKELY to be removed during a transmission removal process?

    A. Torque converter bolts
    B. Alternator
    C. Bell housing bolts
    D. Negative battery cable

39. Technician A says that the transmission needs to be removed to closely inspect the converter flexplate. Technician B says that the starter ring gear should be closely inspected for tooth damage when servicing the flexplate. Who is correct?

    A. A only
    B. B only
    C. Both A and B
    D. Neither A nor B

40. Technician A says that a torque converter that has excessive end-play should be replaced. Technician B says that a torque with a failed one-way clutch should be replaced. Who is correct?

    A. A only
    B. B only
    C. Both A and B
    D. Neither A nor B

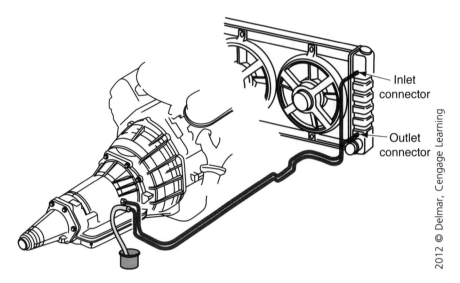

Inlet connector

Outlet connector

2012 © Delmar, Cengage Learning

41. Referring to the figure above, Technician A says the operation in the figure should be performed with the engine at approximately 1,000 rpm. Technician B says that if the volume for this test is less than one quart of fluid in 20 seconds, the transmission cooler should be flushed. Who is correct?

    A. A only

    B. B only

    C. Both A and B

    D. Neither A nor B

42. Which of the following methods would be LEAST LIKELY used in cleaning the mating surface of a valve body and its mating transmission case surface?

    A. Compressed air

    B. Aerosol-based solution

    C. Solvent-based solution

    D. Shop rag

43. Technician A says that an oil pump cover can be checked for flatness with a dial indicator. Technician B says that oil pump gear-to-gear clearance can be checked with a straightedge. Who is correct?

    A. A only

    B. B only

    C. Both A and B

    D. Neither A nor B

44. Technician A says that thrust washers should be measured with a dial caliper to check for correct thickness. Technician B says that some thrust washers are selective and are used to change the end-play on some transmission shafts. Who is correct?

    A. A only

    B. B only

    C. Both A and B

    D. Neither A nor B

45. A technician suspects a piece of debris is lodged in an oil delivery circuit in a transmission case. Which of the following methods would be most likely successful in clearing the debris from this area?

    A. Solvent and compressed air
    B. Pressurized nitrogen
    C. Water
    D. Shop rag and a piece of wire

46. A transmission is being prepared for installation into the car. The new torque converter does not set back in the bell housing as far as it did upon disassembly. Technician A says that this is because the old torque converter was worn. Technician B says that this is because the torque converter is not fully engaged into the pump. Who is correct?

    A. A only
    B. B only
    C. Both A and B
    D. Neither A nor B

47. Technician A says clutch pack clearance can be decreased by adding a thicker shim (reaction/pressure plate). Technician B says clutch pack clearance can be decreased by installing a thicker apply piston. Who is correct?

    A. A only
    B. B only
    C. Both A and B
    D. Neither A nor B

48. Which of the following faults would be LEAST LIKELY to be discovered during a servo air test?

    A. Faulty servo apply pin length
    B. Cracked transmission passage
    C. Servo piston
    D. Servo piston seal

49. Technician A says that the inner race of a sprag clutch should lock in one direction. Technician B says that the inner race of a sprag clutch should freewheel in one direction. Who is correct?

    A. A only
    B. B only
    C. Both A and B
    D. Neither A nor B

50. Technician A says a transmission drum can be damaged by a slipping band. Technician B says that a new band should be soaked in clean transmission fluid before installation. Who is correct?

    A. A only
    B. B only
    C. Both A and B
    D. Neither A nor B

PREPARATION EXAM 2

1. A late-model vehicle needs to be road tested to verify a shifting concern. Technician A says that the shift linkage should be adjusted prior to beginning the road test. Technician B says that the transmission fluid should be checked prior to beginning the road test. Who is correct?

    A. A only
    B. B only
    C. Both A and B
    D. Neither A nor B

2. A damaged transmission mount could cause all of the following problems EXCEPT:

    A. Transmission not shifting into overdrive
    B. Knocking sound when the vehicle is shifted into drive
    C. Knocking sound when the vehicle is shifted into reverse
    D. Vibration during heavy throttle events

3. Which of the following faults would most likely cause a vibration at 40 to 45 mph while turning the steering wheel?

    A. Low line pressure
    B. Slipping clutch pack
    C. Constant velocity (CV) joint
    D. Output shaft seal

4. Technician A says that a faulty clutch pack piston seal can cause a clicking noise when the vehicle is making a turn. Technician B says that a faulty torque converter stator roller clutch can cause a metallic roar while making a turn. Who is correct?

    A. A only
    B. B only
    C. Both A and B
    D. Neither A nor B

5. A vehicle is being diagnosed for a transmission fluid leak near the bell housing area. Technician A says that a failed output shaft seal could cause a leak in this area. Technician B says that a leaking torque converter could cause a leak in this area. Who is correct?

    A. A only
    B. B only
    C. Both A and B
    D. Neither A nor B

6. Technician A says that only the manufacturer-recommended transmission fluid should be added to an automatic transmission. Technician B says that some manufacturers recommend checking the transmission fluid in park. Who is correct?

    A.   A only

    B.   B only

    C.   Both A and B

    D.   Neither A nor B

7. Which gear should a technician put the transmission in to create minimum line pressure?

    A.   Manual second

    B.   Manual low

    C.   Park

    D.   Reverse

8. Technician A says that a pressure test can easily be performed using a scan tool. Technician B says that a pressure test should be performed prior to an initial road test. Who is correct?

    A.   A only

    B.   B only

    C.   Both A and B

    D.   Neither A nor B

9. A vehicle has a stall speed that is 250 rpm below the standard specification. Technician A says that a slipping torque converter stator could be the cause. Technician B says that a plugged fuel filter could be the cause. Who is correct?

    A.   A only

    B.   B only

    C.   Both A and B

    D.   Neither A nor B

10. Technician A says that a stall test should not be performed until the engine warms up. Technician B says that a stall test should be performed for no longer than five seconds. Who is correct?

    A.   A only

    B.   B only

    C.   Both A and B

    D.   Neither A nor B

11. Technician A says that the torque converter lockup clutch is engaged by reversing the fluid flow through the torque converter. Technician B says that the torque converter lockup clutch does not engage at speeds below about 35 mph. Who is correct?

    A.   A only

    B.   B only

    C.   Both A and B

    D.   Neither A nor B

12. A vehicle with an electronically shifted transmission is being road tested. Technician A says that a line pressure gauge can be useful to use during the road test. Technician B says that a shift monitor can be useful during the road test. Who is correct?

    A. A only
    B. B only
    C. Both A and B
    D. Neither A nor B

13. Which of the following tools would most likely be used during the road test of a vehicle with an electronic transmission?

    A. Scan tool
    B. Vacuum gauge
    C. Pressure gauge
    D. Micrometer

14. The pressure control solenoid connector was left disconnected following a transmission installation. This may cause all of the following EXCEPT:

    A. Transmission will jerk when shifting from park to reverse.
    B. Transmission will jerk when shifting from park to drive.
    C. Transmission will slip at every up-shift.
    D. Transmission will have extremely firm up-shifts.

15. Which of the following actions would cause the line pressure to increase in a vehicle with an electronic transmission?

    A. Clearing the trouble codes with a scan tool
    B. Disconnecting the pressure control solenoid
    C. Disconnecting the throttle position sensor
    D. Increasing the amperage signal to the pressure control solenoid with a scan tool

16. A vehicle is being diagnosed for a problem of the engine dying when coming to a stop. After restarting the engine, it again dies when the transmission is shifted into any gear. Technician A says that the torque converter could be the cause. Technician B says that the torque converter clutch solenoid could be the cause. Who is correct?

    A. A only
    B. B only
    C. Both A and B
    D. Neither A nor B

17. Which of the following components would be LEAST LIKELY to cause the torque converter clutch and the cruise control to be inoperative?

    A. Speed sensor
    B. Brake switch
    C. Overdrive switch
    D. Brake light fuse

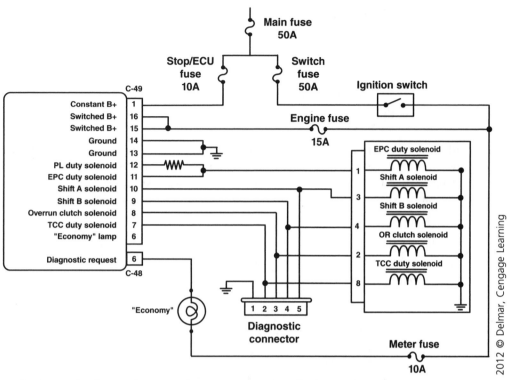

18. Referring to the figure above, Technician A says that all of the solenoids receive ground from the transmission computer. Technician B says that the transmission computer receives power (B+) from the 50 amp main fuse. Who is correct?

A. A only
B. B only
C. Both A and B
D. Neither A nor B

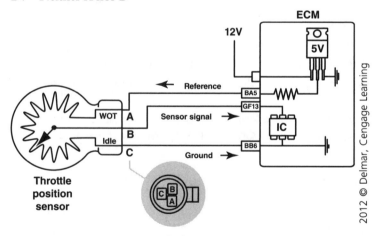

19. Referring to the figure above, Technician A says that the ECM provides a 12 volt supply to the throttle position sensor (TPS). Technician B says that the ECM provides a ground for the TPS at pin BB6 of the ECM. Who is correct?

A. A only
B. B only
C. Both A and B
D. Neither A nor B

20. The wiring and connections of the charging system should be checked in all of the following ways during generator replacement EXCEPT:

    A. Inspect the stator resistance.

    B. Inspect the connections for tightness.

    C. Inspect the wire insulation for cuts and cracks.

    D. Inspect the routing of the wires and harnesses.

21. A battery load test has been performed on an automotive battery. The voltage at the end of the 15-second test was 10.4 volts. Technician A says that the battery should be recharged for 15 minutes and then re-tested. Technician B says that the connectors of the test leads sometimes get warm during this test. Who is correct?

    A. A only

    B. B only

    C. Both A and B

    D. Neither A nor B

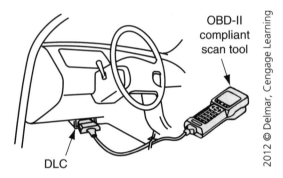

22. The tool shown in the picture above can be used for all of the following purposes EXCEPT:

    A. Accessing the speed sensor coil resistance

    B. Accessing the transmission computer data list

    C. Accessing the diagnostic trouble codes (DTCs) in the engine computer

    D. Accessing the diagnostic trouble codes (DTCs) in the transmission computer

23. A vehicle is being diagnosed for a shuddering condition that is most prevalent when the torque converter clutch engages. Which of the following components would be the most likely cause?

    A. Pressure control solenoid

    B. Torque converter

    C. Modulator

    D. Speed sensor

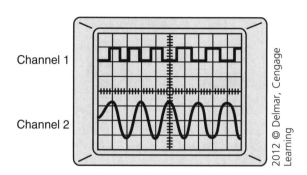

2012 © Delmar, Cengage Learning

24. Referring to the figure above, Technician A says the pattern on Channel 2 of the oscilloscope is a good signal from a permanent magnet generator (speed sensor). Technician B says the pattern from Channel 1 is a good signal from a throttle position sensor. Who is correct?

   A. A only
   B. B only
   C. Both A and B
   D. Neither A nor B

25. Technician A says that a scan tool can be used to retrieve trouble codes from the electronic control module (ECM). Technician B says that a scan tool can be used to retrieve sensor data from the ECM. Who is correct?

   A. A only
   B. B only
   C. Both A and B
   D. Neither A nor B

26. The transmission shift indicator is not aligned when the vehicle is in park. Technician A says that the manual valve linkage may need to be adjusted. Technician B says that the shift indicator could be out of adjustment. Who is correct?

   A. A only
   B. B only
   C. Both A and B
   D. Neither A nor B

27. A transaxle is being diagnosed for a problem of early up-shifts. Which of the following faults would most likely cause this problem?

   A. Broken throttle valve cable
   B. Throttle valve linkage adjusted too tight
   C. Loose band adjustment
   D. Tight band adjustment

28. A transmission is leaking near the rear section of the transmission. Which of the components below is the most likely cause of this leak?

   A. Oil pump seal
   B. Governor cover seal
   C. Manual linkage seal
   D. Input speed sensor seal

29. All of the following conditions could cause repeat output shaft seal failures EXCEPT:

    A. Worn output shaft bushing
    B. Overfilled transmission
    C. Damaged universal joint at the transmission end of the drive shaft
    D. Bent drive shaft

30. Technician A says that the transmission cooler is located in one of the radiator side tanks. Technician B says that leaking transmission cooler lines can be repaired with heater hose and clamps. Who is correct?

    A. A only
    B. B only
    C. Both A and B
    D. Neither A nor B

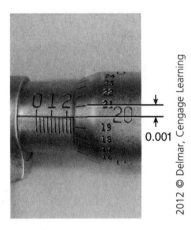

2012 © Delmar, Cengage Learning

31. What is the measurement reading on the (1 to 2 inch) micrometer above?

    A. 1.270 inches
    B. 1.245 inches
    C. 1.220 inches
    D. 1.230 inches

32. A transmission has a broken 1-2 accumulator spring. Technician A says that the transmission will likely shift harshly during the shift from first gear to second gear. Technician B says that the transmission may slip during the shift from second gear to third gear. Who is correct?

    A. A only
    B. B only
    C. Both A and B
    D. Neither A nor B

33. Technician A says that voltage drop testing a connector needs to be done when the circuit is energized. Technician B says that voltage drop testing a relay needs to be done when the circuit is de-energized. Who is correct?

    A. A only
    B. B only
    C. Both A and B
    D. Neither A nor B

34. Technician A states that the scan tool receives data from a connector located on the transmission control module (TCM). Technician B states that the scan tool connects to the data bus using a data link connector (DLC). Who is correct?

    A.   A only
    B.   B only
    C.   Both A and B
    D.   Neither A nor B

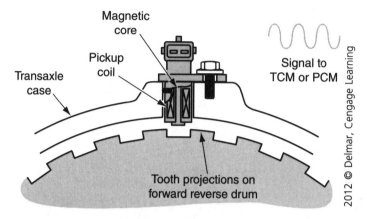

35. Referring to the figure above, which electrical test tool could be used to test the voltage pattern that is shown in the picture?

    A.   Voltmeter
    B.   Oscilloscope
    C.   Ohmmeter
    D.   Continuity tester

36. Which of the following methods of wire repair would be most likely used to resist water intrusion in the repair?

    A.   Twist wires together and wrap with electrical tape.
    B.   Connect wires with Scotch Lock connectors.
    C.   Connect wires by solder and heat shrink.
    D.   Connect wires with butt connectors and wrap with electrical tape.

37. A vehicle is being diagnosed for a bumping noise that occurs when the transmission is shifted to reverse and accelerated. Which of the following components would be the most likely cause?

    A.   Chipped planetary gear
    B.   Broken rear transmission mount
    C.   Transmission band adjusted too loosely
    D.   Slipping clutch pack

38. Which of the following components would be LEAST LIKELY to be removed during a transmission removal process?

    A.   Drive shaft
    B.   Flywheel dust cover
    C.   Rear transmission mount
    D.   Engine flywheel

39. A vehicle has a dark-colored fluid leak around the bell housing area. Faults at which of the following locations would most likely cause this problem?

    A.   Engine rear main seal

    B.   Transmission bottom pan gasket

    C.   Axle seal

    D.   Speed sensor o-ring

40. All of the following flexplate inspections require the transmission to be removed EXCEPT:

    A.   Elongated flexplate mounting holes

    B.   Ring gear

    C.   Loose flexplate bolts

    D.   Cracks in the flexplate

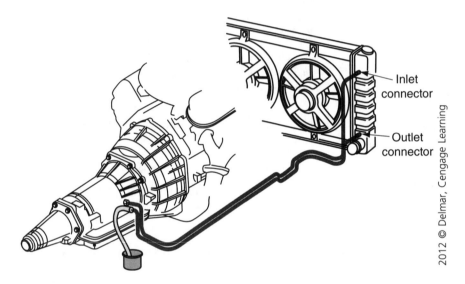

41. Referring to the figure above, Technician A says the operation in the figure should be performed with the engine at 2,500 rpm. Technician B says that the volume for this test should be approximately one quart of fluid in 20 seconds. Who is correct?

    A.   A only

    B.   B only

    C.   Both A and B

    D.   Neither A nor B

42. A clutch pack needs to be disassembled. Technician A says that the clutch pack snap ring should be removed prior to removing the clutch piston. Technician B says that the clutch drum will need to be thoroughly cleaned when all of the internal parts have been removed. Who is correct?

    A. A only
    B. B only
    C. Both A and B
    D. Neither A nor B

43. Technician A says that a new transmission oil pump should be installed dry to prevent contamination. Technician B says that the torque converter hub drives the oil pump on many transmissions. Who is correct?

    A. A only
    B. B only
    C. Both A and B
    D. Neither A nor B

44. Which of the following methods is LEAST LIKELY to be used to adjust end-play on a transmission shaft?

    A. Metal thrust washer
    B. Brass thrust washer
    C. Selective piston plate
    D. Metal spacer shim

45. During a rebuild, many of the transmission components show signs of overheating. Technician A says that all overheated components will need to be replaced. Technician B says that the transmission oil cooler should be inspected for proper operation. Who is correct?

    A. A only
    B. B only
    C. Both A and B
    D. Neither A nor B

46. Technician A says that gasket sealers should be used conservatively to prevent excess sealant from breaking off inside the transmission. Technician B says that some manufacturers do not recommend the use of gasket sealers at all. Who is correct?

    A. A only
    B. B only
    C. Both A and B
    D. Neither A nor B

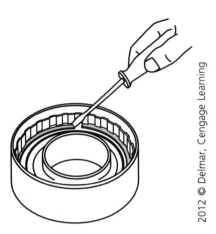

2012 © Delmar, Cengage Learning

47. Which of the following procedures is most likely being performed in the figure above?

   A. Clutch pack apply plate selection

   B. Clutch pack clearance test

   C. Clutch pack snap ring removal

   D. Clutch pack piston return spring service

48. Technician A says the clutch pack clearance may be adjusted by using a selective apply plate. Technician B says that clutch pack clearance may be adjusted by using a selective snap ring. Who is correct?

   A. A only

   B. B only

   C. Both A and B

   D. Neither A nor B

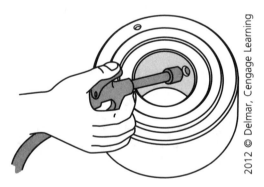

2012 © Delmar, Cengage Learning

49. Referring to the figure above, Technician A says that this test should be performed at each fluid service. Technician B says that this test should be performed at a limited pressure to prevent damage or injury. Who is correct?

   A. A only

   B. B only

   C. Both A and B

   D. Neither A nor B

50. Technician A says that a one-way clutch is energized with hydraulic fluid. Technician B says that a one-way clutch is a holding device. Who is correct?

     A. A only
     B. B only
     C. Both A and B
     D. Neither A nor B

PREPARATION EXAM 3

1. A late-model vehicle needs to be road tested to verify a shifting concern. Technician A says that the road test should be performed under the same conditions as described by the customer. Technician B says that the transmission fluid should be checked prior to beginning the road test. Who is correct?

     A. A only
     B. B only
     C. Both A and B
     D. Neither A nor B

2. Which of the following faults would be LEAST LIKELY to cause a shudder condition at 55 mph during the road test?

     A. Torque converter lockup clutch
     B. Faulty mass airflow sensor
     C. Planetary sun gear chipped
     D. Faulty engine spark plug

3. Technician A says the excessive line pressure can cause harsh shifts. Technician B says that low line pressure can cause a shift that causes a flare in the engine RPM. Who is correct?

     A. A only
     B. B only
     C. Both A and B
     D. Neither A nor B

4. Technician A says that the engine should be shut off when checking the transmission fluid. Technician B says that some transmissions do not have a dipstick to check the transmission fluid. Who is correct?

     A. A only
     B. B only
     C. Both A and B
     D. Neither A nor B

5. A vehicle is being diagnosed that has extremely burned and discolored transmission fluid. Which of the following problems would be LEAST LIKELY to cause this condition with the fluid?

     A. Transmission cooler restricted
     B. Misadjusted band
     C. Faulty transmission computer
     D. Leaking clutch pack piston

6. Technician A says that the governor pressure should be approximately 50 psi at 50 mph. Technician B says that governor pressure should be near zero when the vehicle is stopped. Who is correct?

    A. A only

    B. B only

    C. Both A and B

    D. Neither A nor B

7. A pressure test can reveal all of the following results EXCEPT:

    A. Clutch application pressure

    B. Band pressure

    C. Mainline pressure

    D. Governor pressure

8. A vehicle has a stall speed that is 550 rpm below the standard specification. Technician A says that low fuel pressure to the engine could be the cause. Technician B says that a plugged exhaust system could be the cause. Who is correct?

    A. A only

    B. B only

    C. Both A and B

    D. Neither A nor B

9. Which of the following transmission components is LEAST LIKELY to be tested during a stall test?

    A. Band

    B. Clutch pack

    C. Torque converter stator

    D. Governor valve

10. Technician A says that the engine RPM should drop 25 rpm when the torque converter clutch engages. Technician B says that the torque converter clutch should disengage when the vehicle is slowed down below the engagement speed. Who is correct?

    A. A only

    B. B only

    C. Both A and B

    D. Neither A nor B

11. Technician A says that the torque converter lockup clutch is not typically functional on a cold engine. Technician B says that the torque converter lockup clutch creates high levels of heat in the torque converter. Who is correct?

    A. A only

    B. B only

    C. Both A and B

    D. Neither A nor B

12. Which of the following tools would most likely be used during the road test of a vehicle with an electronic transmission?

    A.   Dial indicator
    B.   Vacuum gauge
    C.   Pressure gauge
    D.   Scan tool

13. A vehicle with an electronic transmission is being road tested. All of the following items should be observed during the road test EXCEPT:

    A.   Correct up-shift speeds
    B.   Unusual drive train noises
    C.   Unusual downshift vibration
    D.   Fuel economy

14. A vehicle with an electronic pressure control solenoid is being diagnosed. Technician A says that maximum pressure can be commanded by using a scan tool in the actuator mode. Technician B says that maximum pressure can be attained by putting the transmission in neutral. Who is correct?

    A.   A only
    B.   B only
    C.   Both A and B
    D.   Neither A nor B

15. A vehicle with an electronic transmission has been tested for transmission pump output pressure. The pressure test results are below the specification in all ranges. Which of the following conditions would most likely cause this problem?

    A.   Slipping torque converter stator
    B.   Front pump seal leaking externally
    C.   Loose torque converter mounting bolt
    D.   Transmission filter not properly secured to the pump

16. A vehicle is being diagnosed for a problem of the engine dying when coming to a stop. After restarting the engine, it again dies when the transmission is shifted into any gear. Which of the following is most likely to be the cause?

    A.   An open transmission temperature sensor
    B.   A shorted speed sensor
    C.   A faulty generator
    D.   Stuck torque converter clutch solenoid

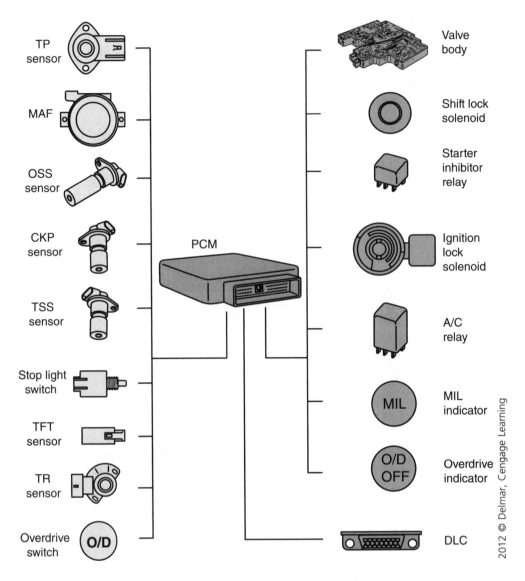

17. Referring to the figure above, Technician A says that the PCM controls the engine and the electronic transmission. Technician B says that the PCM does not control the shift interlock system. Who is correct?

    A. A only
    B. B only
    C. Both A and B
    D. Neither A nor B

18. Technician A says that a failed brake switch could cause the torque converter clutch to be inoperative. Technician B says that a stopped-up transmission filter could cause the torque converter clutch to be inoperative. Who is correct?

    A. A only
    B. B only
    C. Both A and B
    D. Neither A nor B

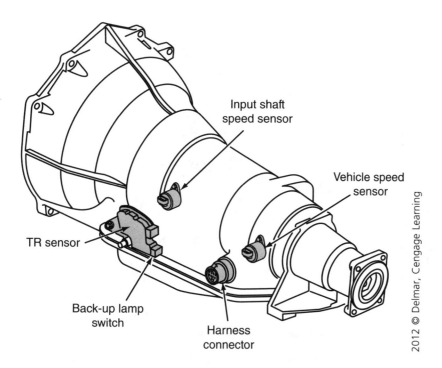

Input shaft speed sensor

Vehicle speed sensor

TR sensor

Back-up lamp switch

Harness connector

2012 © Delmar, Cengage Learning

19. Referring to the figure above, Technician A says that the input shaft speed sensor is a thermistor that varies its resistance as the vehicle speed changes. Technician B says that the input shaft speed sensor is used on electronic transmissions as an input to calculate the gear ratio to determine when a clutch pack may be slipping. Who is correct?

A. A only

B. B only

C. Both A and B

D. Neither A nor B

20. A digital battery tester would be most likely used for which of the following purposes?

A. Measuring resistance of control modules

B. Testing starter current draw

C. Measuring voltage of control modules

D. Testing the impedance of the battery

21. Technician A says that the voltage drop on the positive battery cable should be less than 0.5 volts while cranking the engine. Technician B says that the voltage drop on the negative battery cable should be less than 1.5 volts while cranking the engine. Who is correct?

A. A only

B. B only

C. Both A and B

D. Neither A nor B

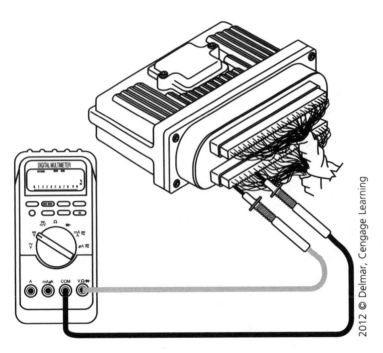

2012 © Delmar, Cengage Learning

22. Referring to the figure above, Technician A says that the meter is measuring the resistance of the module. Technician B says that all meters used in this manner need to be a high-impedance design. Who is correct?

    A. A only
    B. B only
    C. Both A and B
    D. Neither A nor B

23. A vehicle is being diagnosed for a shuddering condition that is most prevalent when the torque converter clutch engages. This problem could be caused by all of the following components EXCEPT:

    A. Pressure control solenoid
    B. Pulse-width modulated torque converter clutch solenoid
    C. Torque converter clutch apply plate
    D. Torque converter clutch friction disc

24. A problem with the following electronic transmission components could cause the transmission to enter "limp mode" EXCEPT:

    A. Shift solenoid A
    B. Overdrive switch
    C. Shift solenoid B
    D. Input shaft speed sensor

25. A vehicle with an electronic transmission will not move in forward or reverse. Which of the following could cause this problem?

    A. Broken input shaft
    B. Faulty input shaft speed sensor
    C. Faulty shift solenoid
    D. Transmission fluid level is one quart overfilled

26. Which of the components below is used as an input to the transmission computer for gearshift position?

    A. MAP sensor
    B. Back-up switch
    C. Range sensor
    D. Manual valve

27. Technician A says that a loosely adjusted throttle valve cable will cause early up-shifts. Technician B says that a throttle valve cable adjusted too tight will cause late up-shifts. Who is correct?

    A. A only
    B. B only
    C. Both A and B
    D. Neither A nor B

28. A transmission is leaking near the rear section of the transmission. All of the components below could cause this problem EXCEPT:

    A. Oil pump seal
    B. Extension housing gasket
    C. Output shaft seal
    D. Output speed sensor seal

29. A transmission is being diagnosed for a worn extension housing bushing. Technician A says that the old bushing could be compared to the new bushing during the inspection. Technician B says that the drive yoke should be closely checked for deep scratches during the inspection. Who is correct?

    A. A only
    B. B only
    C. Both A and B
    D. Neither A nor B

30. A vehicle is in the shop for an overheating problem. During the inspection the technician finds that upper radiator hose collapses after the engine cools down but moves back to normal when the radiator cap is removed. What is the most likely cause for this problem?

    A. Faulty water pump
    B. Malfunctioning cooling fan
    C. Faulty radiator cap
    D. Faulty thermostat

31. Technician A says that the valve body is typically tightened up with an inch/pound torque wrench. Technician B says that the valve body should be tightened to a specified torque in a uniform manner. Who is correct?

    A. A only
    B. B only
    C. Both A and B
    D. Neither A nor B

32. Technician A says that intermittent electrical signals can sometimes be diagnosed by using an oscilloscope. Technician B says that a broken tooth on a speed sensor reluctor can be sometimes be diagnosed by using an ohmmeter. Who is correct?

    A. A only
    B. B only
    C. Both A and B
    D. Neither A nor B

33. Which electrical tool would be LEAST LIKELY to be used when testing transmission sensors?

    A. Digital voltmeter
    B. Oscilloscope
    C. Test light
    D. Digital ohmmeter

34. Which of the following definitions best describes a terminal?

    A. A magnetic switch
    B. A metallic component used to tie electrical circuits together
    C. A plastic housing used to hold terminals as well as to plug into electrical items
    D. A joint where two or more electrical circuits connect to each other

35. Which of the following conditions would be the LEAST LIKELY cause of a thumping sound that happens when accelerating quickly?

    A. Broken engine mount
    B. Faulty axle bearing
    C. Broken transmission mount
    D. Faulty universal joint

36. Which gear is most likely to be selected when checking the transmission fluid level?

    A. Manual low
    B. Park
    C. Reverse
    D. Overdrive

37. Which type of transmission fluid should be used in an automatic transmission?

    A. Dexron 4
    B. Manufacturer recommended type
    C. ATF+4
    D. Dexron with Mercon

38. A transaxle needs to be removed from a late-model car. Technician A installs an engine support fixture before removing the engine subframe. Technician B removes the air cleaner before removing the upper bell housing bolts. Who is correct?

    A. A only
    B. B only
    C. Both A and B
    D. Neither A nor B

39. A transaxle is ready for installation. Technician A inspects the torque converter closely to make sure that it is completely engaged into the front of the transaxle. Technician B inspects the engine dowel pins to make sure that they are in place. Who is correct?

    A.   A only
    B.   B only
    C.   Both A and B
    D.   Neither A nor B

40. Technician A says that the torque converter position should be noted prior to removing from the transmission. Technician B says that the torque converter should be thoroughly installed into the transmission prior to installing the transmission. Who is correct?

    A.   A only
    B.   B only
    C.   Both A and B
    D.   Neither A nor B

41. Technician A says that an external transmission cooler should be used on any vehicle that is loaded heavily on a regular basis. Technician B says that some transmission coolers have small baffles to increase the heat transfer. Who is correct?

    A.   A only
    B.   B only
    C.   Both A and B
    D.   Neither A nor B

42. Which of the following methods would be most likely used in cleaning the transmission case after the internal parts have been removed?

    A.   Parts solvent and compressed air
    B.   Air-powered die grinder with a buffing wheel
    C.   Aerosol cleaner and a paper shop towel
    D.   Parts solvent and a shop rag

43. Which of the following tools would be LEAST LIKELY to be used when measuring a gear-type oil pump assembly?

    A.   Feeler gauge
    B.   Straightedge
    C.   Ruler
    D.   Micrometer

44. Technician A says that bearing preload is sometimes measured by testing the turning torque on a shaft that has been assembled. Technician B says that backlash is measured with a micrometer. Who is correct?

    A.   A only
    B.   B only
    C.   Both A and B
    D.   Neither A nor B

45. A plastic thrust washer is found to be embedded with metal particles during transmission disassembly. Technician A says that any plastic thrust washers with metal embedded in them should be replaced. Technician B says that the plastic washers should be replaced with metal washers to prevent this from happening again. Who is correct?

    A. A only

    B. B only

    C. Both A and B

    D. Neither A nor B

46. Technician A says that transmission shafts should be checked for scoring on all areas that ride next to a bushing. Technician B says that transmission shafts should be checked for scoring on all areas that ride next to a bearing. Who is correct?

    A. A only

    B. B only

    C. Both A and B

    D. Neither A nor B

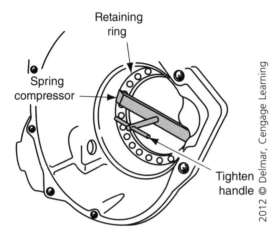

47. Which of the following procedures is most likely being performed in the figure above?

    A. Clutch pack apply plate selection

    B. Clutch pack clearance test

    C. Clutch pack snap ring removal

    D. Clutch pack piston return spring service

48. All of the following methods are sometimes used to adjust the clutch pack clearance EXCEPT:

    A. Selective piston assembly

    B. Selective apply plate

    C. Adding an extra steel plate

    D. Selective snap ring

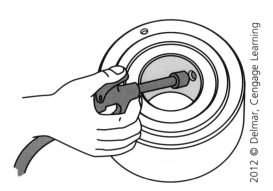

49. Referring to the figure above, Technician A says that this test should be performed prior to assembling the clutch pack into the transmission. Technician B says that this test will expose a problem in the clutch pack piston and seal area. Who is correct?

    A. A only
    B. B only
    C. Both A and B
    D. Neither A nor B

50. A transmission is being diagnosed for a problem of slipping in "OD" first gear. However, the transmission does not slip when the manual low position is used on the gearshift. What is the most likely cause of this problem?

    A. The torque converter impeller is weak.
    B. The low gear roller clutch is slipping.
    C. The forward clutch pack assembly has a blown piston seal.
    D. The low gear servo is leaking.

PREPARATION EXAM 4

1. Technician A says that a broken transmission mount assembly could cause a knocking noise when the vehicle is accelerated quickly during a road test. Technician B says that a missing retaining bolt for a transmission mount could cause a knocking noise when the vehicle is shifted into reverse during a road test. Who is correct?

    A. A only
    B. B only
    C. Both A and B
    D. Neither A nor B

2. During a road test, the technician notices that the transmission slips when making a sharp turn. Technician A says that the torque converter clutch could be staying engaged too long. Technician B says that the speed sensor could be dropping out when the vehicle is making a turn. Who is correct?

    A. A only
    B. B only
    C. Both A and B
    D. Neither A nor B

3. Which of the following drive train components would be most likely to cause a roaring noise that increases when the wheels are turned to the right?

   A. Right-front wheel bearing
   B. Loose torque converter bolt
   C. Transmission oil pump
   D. Left-front wheel bearing

4. All of the following faults could cause a clicking noise when the vehicle is put into gear and quickly accelerated while holding the brakes EXCEPT:

   A. Worn spider gears
   B. Cracked flexplate
   C. Loose torque converter bolts
   D. Loose flywheel bolts

5. A vehicle is being diagnosed for a transmission fluid leak near the bell housing area. Technician A says that a failed output shaft seal could cause a leak in this area. Technician B says that a leaking torque converter could cause a leak in this area. Who is correct?

   A. A only
   B. B only
   C. Both A and B
   D. Neither A nor B

6. All of the following problems could cause low transmission fluid level EXCEPT:

   A. Leaking transmission pan gasket
   B. Cracked valve body channel plate
   C. Faulty modulator valve
   D. Output shaft seal

7. Technician A says that the governor pressure should be approximately 30 psi at 15 mph. Technician B says that governor pressure should be near zero when the vehicle is parked. Who is correct?

   A. A only
   B. B only
   C. Both A and B
   D. Neither A nor B

8. Which of the following components is LEAST LIKELY to have an effect on line pressure?

   A. Governor valve
   B. Manual valve
   C. Throttle valve
   D. Gear selector

9. A vehicle has a stall speed that is 325 rpm below the standard specification. Technician A says that a locked torque converter stator could be the cause. Technician B says that a weak transmission oil pump could be the cause. Who is correct?

   A. A only
   B. B only
   C. Both A and B
   D. Neither A nor B

10. All of the following transmission components are being tested during a stall test EXCEPT:

    A. Oil pump capacity
    B. Band holding capacity
    C. Clutch pack holding capacity
    D. Torque converter stator

11. Technician A says that the torque converter lockup clutch may be controlled by a lockup solenoid. Technician B says that the torque converter lockup clutch does not engage at speeds below about 55 miles per hour. Who is correct?

    A. A only
    B. B only
    C. Both A and B
    D. Neither A nor B

12. Technician A says that the torque converter lockup clutch can be commanded to turn on by using a bi-directional scan tool. Technician B says that the engine RPM should increase approximately 150 rpm when the converter lockup clutch engages. Who is correct?

    A. A only
    B. B only
    C. Both A and B
    D. Neither A nor B

13. A vehicle with an electronic transmission is being diagnosed. Technician A says that it is wise to perform a line pressure test prior to conducting a road test. Technician B says that it is wise to check the transmission fluid prior to conducting a road test. Who is correct?

    A. A only
    B. B only
    C. Both A and B
    D. Neither A nor B

14. A vehicle with an electronic transmission is being road tested. All of the following items should be observed during the road test EXCEPT:

    A.   Torque converter lockup clutch engagement

    B.   Torque converter lockup clutch disengagement

    C.   Engine oil pressure

    D.   Up-shift smoothness

15. The pressure control solenoid coil has an open circuit. Technician A says that the transmission will likely jerk when it is shifted from park to reverse. Technician B says that the transmission will likely still have smooth up-shifts. Who is correct?

    A.   A only

    B.   B only

    C.   Both A and B

    D.   Neither A nor B

16. A vehicle with an electronic transmission has been tested for transmission pump output pressure. The pressure test results are below the specification in all ranges. All of the following conditions could cause this problem EXCEPT:

    A.   Restricted transmission filter

    B.   Slipping clutch pack

    C.   Punctured transmission fluid pickup tube

    D.   Excess transmission pump clearance

17. All of the following conditions will cause the torque converter clutch to disengage EXCEPT:

    A.   Applying the brakes

    B.   Heavy throttle application

    C.   Decreasing vehicle speed below minimum engagement specification

    D.   Shifting the gear shift into overdrive

18. A vehicle is being diagnosed for a problem with the torque converter clutch system. Technician A says that the torque converter clutch should be engaged at 45 mph after the engine has warmed up. Technician B says that the torque converter clutch may disengage when the vehicle climbs a steep hill at highway speed. Who is correct?

    A.   A only

    B.   B only

    C.   Both A and B

    D.   Neither A nor B

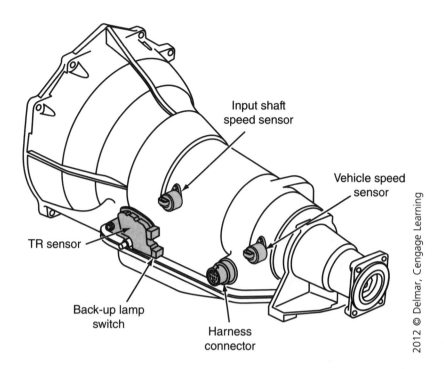

19. Referring to the figure above, Technician A says that the TR sensor is used to signal the selected range to the PCM. Technician B says that the harness connector can be used to test the resistance of the internal shift solenoids with an ohmmeter. Who is correct?

    A. A only
    B. B only
    C. Both A and B
    D. Neither A nor B

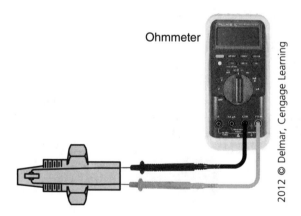

20. Referring to the figure above, Technician A says that the fluid temperature sensor resistance should decrease as the temperature of the sensor increases. Technician B says that the fluid temperature sensor resistance should increase as the temperature decreases. Who is correct?

    A. A only
    B. B only
    C. Both A and B
    D. Neither A nor B

21. The battery housing received some damage from driving a vehicle on rough roads. Electrolyte spilled all over the battery tray. Technician A says that brake cleaner should be used to clean the area. Technician B says that baking soda could be used to neutralize the battery acid. Who is correct?

    A. A only
    B. B only
    C. Both A and B
    D. Neither A nor B

22. All of the following conditions could cause low charging system output EXCEPT:

    A. Loose starter mounting bolts
    B. Loose terminal connection at the charging system "Maxi Fuse"
    C. Loose generator mounting bracket
    D. Damaged charging output wire

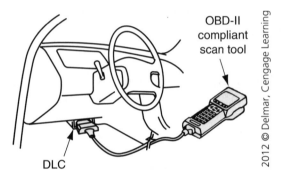

23. The tool shown in the picture above can be used for which of the following purposes?

    A. Testing the vehicle battery
    B. Checking the charging system output
    C. Accessing the transmission computer data list
    D. Checking for the supply voltage (B+) at the transmission computer

24. A vehicle has low power when taking off from a stop, but the power seems normal at highway speeds. Technician A says that the engine exhaust system could be restricted. Technician B says that the torque converter stator could be defective. Who is correct?

    A. A only
    B. B only
    C. Both A and B
    D. Neither A nor B

25. A problem with which of the following electronic transmission components would be most likely to cause the transmission to have extremely firm shifts?

    A. Range sensor
    B. Output speed sensor
    C. Input shaft speed sensor
    D. Pressure control solenoid

26. The main power relay for the electronic control system has failed. Technician A says that the vehicle will not move in forward or reverse. Technician B says that this fault should set a diagnostic trouble code (DTC) in the transmission computer. Who is correct?

    A. A only
    B. B only
    C. Both A and B
    D. Neither A nor B

27. Technician A says that a misadjusted range sensor could cause a no-crank condition. Technician B says that a misadjusted park/neutral switch could cause a no-crank condition. Who is correct?

    A. A only
    B. B only
    C. Both A and B
    D. Neither A nor B

28. A transmission will move in forward and reverse but has no up-shifts. Technician A says that a broken kick-down cable could be the cause. Technician B says that a faulty governor valve could be the cause. Who is correct?

    A. A only
    B. B only
    C. Both A and B
    D. Neither A nor B

29. A transmission needs to have an extension housing gasket replaced. Technician A says that the drive shaft will need to be removed during this process. Technician B says that the transmission will have to be removed from the vehicle to perform this repair on most transmissions. Who is correct?

    A. A only
    B. B only
    C. Both A and B
    D. Neither A nor B

30. Technician A says that the transmission vent should be checked when a repetitive seal failure occurs. Technician B says that the drive shaft has to be removed to replace the extension housing output seal. Who is correct?

    A. A only
    B. B only
    C. Both A and B
    D. Neither A nor B

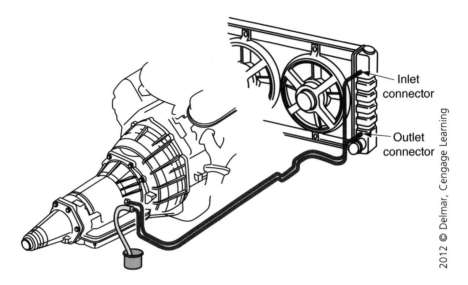

2012 © Delmar, Cengage Learning

31. Referring to the figure above, Technician A says that the cooler flow test should be performed at each transmission service. Technician B says that the radiator may have to be replaced if the amount measured is less than one quart in 20 seconds during the test. Who is correct?

    A. A only

    B. B only

    C. Both A and B

    D. Neither A nor B

32. Technician A says that valve body bolts are often tightened using the torque-to-yield method. Technician B says that the valve body bolts should be marked as they are removed to assure that they get reinstalled in the correct location. Who is correct?

    A. A only

    B. B only

    C. Both A and B

    D. Neither A nor B

33. A transmission has a broken 1-2 accumulator spring. Technician A says that the transmission will likely slip during the first-to-second-gear shift sequence. Technician B says that the line pressure will likely increase due to the broken spring. Who is correct?

    A. A only

    B. B only

    C. Both A and B

    D. Neither A nor B

34. Which of the following tests can be performed with a fused jumper wire?

    A. By-passing a rheostat

    B. By-passing a thermistor

    C. By-passing the load side of a relay

    D. By-passing a potentiometer

35. Technician A says that a test light can be used to test for voltage on computer data circuits. Technician B says that a test light can be used to check for power on the input and output sides of fuses at the fuse panel. Who is correct?

    A. A only
    B. B only
    C. Both A and B
    D. Neither A nor B

36. Which of the following definitions best describes a relay?

    A. A magnetic switch
    B. A metallic component that is used to tie electrical circuits together
    C. A plastic housing that is used to hold terminals, as well as to plug into electrical items
    D. A joint where two or more electrical circuits connect to each other

37. A vehicle is being diagnosed for a problem of the accelerator pedal stuck in the wide-open throttle position. Which of the following components could cause this condition?

    A. Broken cruise control cable
    B. Broken throttle cable
    C. Broken engine mount
    D. Shorted throttle position sensor

38. Where is the transmission filter located on an automatic transmission?

    A. Inside the transmission gear train
    B. Inside the transmission pan
    C. Inside the transmission cooler
    D. Inside the torque converter

39. A vehicle has a fluid leak around the bell housing area. Which of the following faults would be LEAST LIKELY to cause this problem?

    A. Transmission front pump seal
    B. Transmission bottom pan gasket
    C. Engine rear main seal
    D. Transmission front pump retaining bolts

40. A transmission is ready for installation. Which of the following inspections would be LEAST LIKELY to be needed prior to this installation?

    A. Crankshaft pilot bore
    B. Cracks in flexplate
    C. Clutch pack clearance
    D. Elongated mounting holes in flexplate

41. Technician A says that the transmission cooler should be flushed every time that the transmission fluid is serviced. Technician B says that the cooler should be reverse flushed if possible to increase the chances of a thorough cleaning. Who is correct?

    A. A only
    B. B only
    C. Both A and B
    D. Neither A nor B

42. Which component has to be replaced in order to replace the primary transmission fluid cooler?

    A. Condenser
    B. Reserve tank
    C. Radiator
    D. Water pump

43. Which of the following statements about bearing preload is LEAST LIKELY to be correct?

    A. A bearing preload that is set too tight will loosen up once the unit heats up.
    B. A torque wrench is needed to set the bearing preload.
    C. Bearing preload is sometimes checked by testing the turning effort of a component.
    D. Bearings must be lubricated in order to perform under an extended load.

44. Technician A says that a clean shop rag should be used to clean an internal oil passage on a transmission shaft. Technician B says that solvent could be used to internally clean an internal oil passage on a transmission shaft. Who is correct?

    A. A only
    B. B only
    C. Both A and B
    D. Neither A nor B

45. Which of the following actions would be LEAST LIKELY performed when inspecting a planetary gear set?

    A. Measure the pinion gear end-play with a feeler gauge.
    B. Measure the bearing preload of the pinion gears.
    C. Inspect the pinion gears for damaged teeth.
    D. Measure the pinion gear end-play with dial indicator.

46. A mating surface for the valve body is found to be warped. Technician A says that the transmission will have to be machined in order to correct the warpage. Technician B says that a special spacer plate may be used to accommodate the uneven surface. Who is correct?

    A. A only
    B. B only
    C. Both A and B
    D. Neither A nor B

47. Which of the following tools would most likely be used when measuring the backlash on the final drive gears in a transaxle?

    A. Dial indicator
    B. Slide caliper
    C. Outside micrometer
    D. Feeler gauge

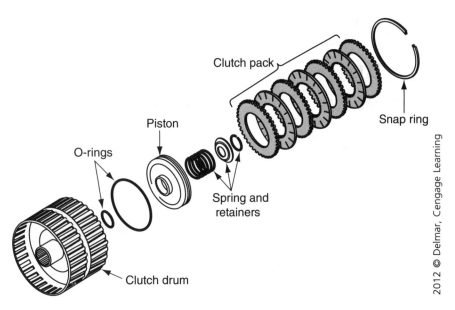

48. What is the purpose of the spring shown in the figure above?

    A. To apply the piston when fluid stops flowing into the clutch pack
    B. To release the piston when fluid stops flowing into clutch pack
    C. To hold pressure on the clutch pack when fluid is sent to the assembly
    D. To release pressure on the clutch pack when fluid is sent to the assembly

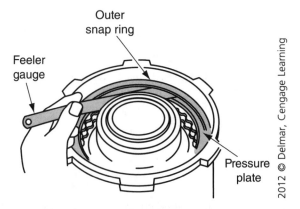

49. Which of the follow procedures is most likely being performed in the figure above?

    A. Clutch pack apply plate selection
    B. Clutch pack clearance test
    C. Clutch pack snap ring removal
    D. Clutch pack piston return spring service

50. Technician A says that air testing a servo assembly will reveal a faulty servo piston seal. Technician B says that air testing a servo assembly will reveal a faulty servo piston return spring. Who is correct?

    A. A only
    B. B only
    C. Both A and B
    D. Neither A nor B

PREPARATION EXAM 5

1. During a road test, the technician notices that the transmission slips when making a sharp turn. Technician A says that the pressure control solenoid could be sticking. Technician B says that the transmission fluid level could be low. Who is correct?

    A. A only

    B. B only

    C. Both A and B

    D. Neither A nor B

2. A technician notices that the engine exhaust system is emitting white smoke during a road test. Which of the following transmission components could cause this condition?

    A. Misadjusted throttle valve cable

    B. Ruptured vacuum modulator

    C. Plugged transmission cooler

    D. Faulty manifolf absolute pressure (MAP) sensor

3. Which of the following faults would be most likely to cause a clicking noise that increases when the transmission is shifted into drive and goes away when the transmission is shifted into park?

    A. Worn transmission pump

    B. Loose torque converter bolts

    C. Slipping transmission clutch pack

    D. Torque converter lockup clutch slipping

4. A vehicle is being diagnosed for a power train vibration at 50 to 60 mph. Technician A says that a bent drive shaft could be the cause. Technician B says that a faulty bent pinion yoke could be the problem. Who is correct?

    A. A only

    B. B only

    C. Both A and B

    D. Neither A nor B

5. Technician A says that the engine should be running when checking the transmission fluid. Technician B says that the dipstick should reveal which gear to put the transmission in to check the transmission fluid. Who is correct?

    A. A only

    B. B only

    C. Both A and B

    D. Neither A nor B

6. A transmission pressure test has been performed on a vehicle. Which of the following pieces of data would be LEAST LIKELY revealed from this test?

    A. Clutch return spring pressure

    B. Servo application pressure

    C. Mainline pressure

    D. Governor pressure

7. A vehicle will back-up and go into first gear without a problem. The vehicle will not up-shift until very high engine RPM. Which of the following conditions would most likely cause the problem?

    A. Manual valve is out of adjustment.
    B. Transmission fluid is overfilled.
    C. Band is out of adjustment.
    D. Governor valve is sticking.

8. A vehicle has a stall speed that is 25 rpm below the standard specification. Technician A says that a locked torque converter stator could be the cause. Technician B says that this is within the normal specifications. Who is correct?

    A. A only
    B. B only
    C. Both A and B
    D. Neither A nor B

9. Technician A says that the stall test should not be continued for longer than five seconds. Technician B says that performing stall tests increases the temperature of the transmission fluid greatly. Who is correct?

    A. A only
    B. B only
    C. Both A and B
    D. Neither A nor B

10. Technician A says that some transmission computers will engage the torque converter lockup clutch under heavy loads when the transmission fluid temperature rises above safe levels. Technician B says that the torque converter lockup clutch is locked up when the pump and turbine are near the same RPM. Who is correct?

    A. A only
    B. B only
    C. Both A and B
    D. Neither A nor B

11. Technician A says that the flow to the transmission cooler is restricted when the torque converter lockup clutch is disengaged. Technician B says that the torque converter builds high levels of heat when the lockup clutch is engaged. Who is correct?

    A. A only
    B. B only
    C. Both A and B
    D. Neither A nor B

12. A vehicle with an electronic transmission is being diagnosed. Technician A says that a scan tool can expose possible problems during the road test. Technician B says that it is a good idea to check the transmission fluid prior to conducting a road test. Who is correct?

    A. A only
    B. B only
    C. Both A and B
    D. Neither A nor B

13. Which of the following items should be monitored during the road test of a vehicle with an electronic transmission?

    A. Transmission fluid color

    B. Transmission cooler flow rate

    C. Drive train noises

    D. Transmission fluid level

14. All of the following actions would cause transmission line pressure to rapidly increase EXCEPT:

    A. Disconnecting the transmission fluid temperature (TFT) sensor

    B. Using a scan tool to command the pressure control solenoid to increase pressure

    C. Disconnecting the pressure control solenoid

    D. Shifting the transmission to reverse and moving the throttle to wide-open throttle (WOT)

15. A vehicle with an electronic pressure control solenoid is being diagnosed. Technician A says that that maximum pressure can be attained by putting the transmission in reverse and moving the throttle to wide-open throttle (WOT). Technician B says that maximum pressure can be attained by using a scan tool and commanding the pressure control solenoid to receive minimum amperage. Who is correct?

    A. A only

    B. B only

    C. Both A and B

    D. Neither A nor B

16. Which of the following components could cause the torque converter clutch and the cruise control to be inoperative?

    A. Cruise control switch

    B. Brake switch

    C. Overdrive switch

    D. Headlight switch

17. Which of the following conditions would be LEAST LIKELY to cause a drop in fuel economy?

    A. Coolant temperature sensor shorted out

    B. Torque converter clutch inoperative

    C. Engine thermostat stuck open

    D. Worn transmission oil pump

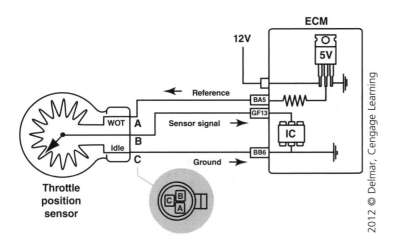

18. Referring to the figure above, which terminal would need to be back-probed to test the signal voltage for the throttle position sensor?

   A. Terminal A of the throttle position sensor

   B. Terminal BA5 of the ECM

   C. Terminal B of the throttle position sensor

   D. Terminal C of the throttle position sensor

19. All of the following statement are true concerning technical service bulletins EXCEPT:

   A. Technical service bulletins are created to provide free repair services to vehicles that exhibit repeat failures.

   B. Technical service bulletins are issued by the vehicle manufacturer.

   C. Technical service bulletins are issued to assist technicians in repairing pattern failures.

   D. Technical service bulletins are available to aftermarket technicians who have professional databases.

20. A late-model vehicle is being diagnosed for a charging problem. The generator only charges at 12.4 volts. A voltage drop test is performed on the charging output wire and 1.8 volts is measured. Technician A says that a blown fusible link in the output circuit could be the cause. Technician B says that a loose nut at the charging output connector could be the cause. Who is correct?

   A. A only

   B. B only

   C. Both A and B

   D. Neither A nor B

21. A maintenance-free battery is low on electrolyte. Technician A says a defective voltage regulator may cause this problem. Technician B says a loose alternator belt may cause this problem. Who is correct?

   A. A only

   B. B only

   C. Both A and B

   D. Neither A nor B

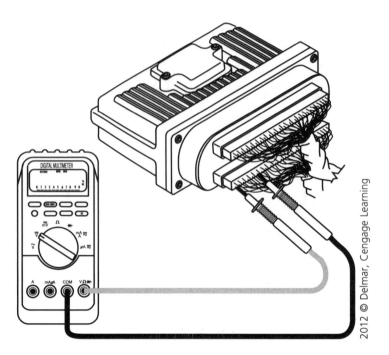

2012 © Delmar, Cengage Learning

22. Referring to the figure above, Technician A says that the meter leads are back-probing the module connector. Technician B says that all meters used in this manner need to be a high-impedance design. Who is correct?

   A. A only

   B. B only

   C. Both A and B

   D. Neither A nor B

23. A vehicle has low power at all speeds. Technician A says that the engine exhaust system could be restricted. Technician B says that the torque converter stator could be defective. Who is correct?

   A. A only

   B. B only

   C. Both A and B

   D. Neither A nor B

24. A vehicle with an electronic transmission is being diagnosed for a problem that it stays in second gear all of the time. Technician A says that a scan tool can be used to determine if the transmission computer has any diagnostic trouble codes (DTCs). Technician B says a scan tool can sometimes be used to override the transmission computer and command an up-shift while in the output test mode. Who is correct?

   A. A only

   B. B only

   C. Both A and B

   D. Neither A nor B

25. A vehicle with an electronic transmission is being diagnosed for only having one forward gear and reverse gear. Technician A says that dirty transmission fluid could be the cause. Technician B says that a restricted cooler return pipe could be the cause. Who is correct?

    A. A only

    B. B only

    C. Both A and B

    D. Neither A nor B

26. Technician A says that a misadjusted shift linkage could cause a no-crank condition. Technician B says that a range switch could cause a no-crank condition. Who is correct?

    A. A only

    B. B only

    C. Both A and B

    D. Neither A nor B

27. Technician A says that the kick-down cable can be replaced without removing the transmission pan. Technician B says that the kick-down cable is not usually adjustable. Who is correct?

    A. A only

    B. B only

    C. Both A and B

    D. Neither A nor B

28. Transmission fluid is leaking from the input speed sensor area. Technician A says that the speed sensor o-ring can be replaced. Technician B says that the leak may be a cracked speed sensor. Who is correct?

    A. A only

    B. B only

    C. Both A and B

    D. Neither A nor B

29. Which of the following cooling system tests would be LEAST LIKELY to be performed on a late-model vehicle?

    A. Pressure test for leaks

    B. Freeze protection test with a refractometer

    C. Hose inspection

    D. Leak detection test with an electronic leak detector

30. Technician A says that the valve body check balls should be cleaned with a mineral spirits and compressed air. Technician B says that the spool valves should be cleaned with light sandpaper during an overhaul procedure. Who is correct?

    A. A only

    B. B only

    C. Both A and B

    D. Neither A nor B

31. All of the following statements about the engine cooling system are correct EXCEPT:

    A. The boiling point is increased when the system builds pressure.

    B. The boiling point is decreased when the coolant is mixed with the water.

    C. The coolant recovery tank has a "full hot" and a "full cold" level.

    D. The freezing point is lowered when the coolant is mixed with the water.

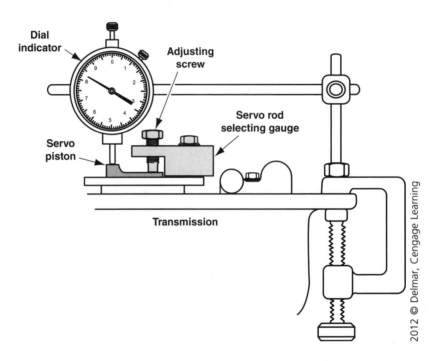

32. What process is taking place in the figure above?

    A. Testing the servo piston end-play

    B. Testing the servo rod end-play

    C. Selecting the correct servo apply pin

    D. Selecting the correct servo piston

33. What will be the most likely transmission condition if the servo piston fails to apply?

    A. The transmission will shift harshly in the gear that the servo is applied in.

    B. The transmission will likely slip in the gear that the band is supposed to be applied in.

    C. The vehicle will not move in forward or reverse gear.

    D. The transmission will flare at the time that the band is commanded to be applied.

34. Which of the following methods of wire repair would be most likely used to resist water intrusion in the repair?

    A. Twist wires together and wrap with electrical tape.

    B. Connect wires with Scotch Lock connectors.

    C. Connect wires with a crimp-and-seal connector and heat with a heat gun.

    D. Connect wires with butt connectors and wrap with electrical tape.

35. Technician A says that a relay is used to control a high-level current circuit by using a small-current circuit. Technician B says that solenoids are used in electronic transmissions to control the transmission fluid pressure. Who is correct?

    A. A only

    B. B only

    C. Both A and B

    D. Neither A nor B

36. Which method would be most likely used when testing for live circuit voltage at a connector?

    A. Piercing the wire insulation with a special tool

    B. Back-probing the connector with a T pin

    C. Disconnecting the connector to front-probe the connection

    D. Disconnecting the connector to back-probe the connection with a T pin

37. Which method would be used to replace all of the transmission fluid?

    A. Transmission drain plug removal

    B. Bottom transmission pan removal

    C. Fluid exchange machine

    D. Cooler line removal

38. Technician A says that an engine support fixture should be installed on a vehicle prior to removing rear-wheel drive transmission. Technician B says that the drive shaft will have to be removed during the process of removing the transmission. Who is correct?

    A. A only

    B. B only

    C. Both A and B

    D. Neither A nor B

39. Technician A says that the torque converter can be easily forced into the transmission with the bell housing bolts while installing the transmission. Technician B says that the converter pilot should perfectly align with the crankshaft pilot bore during installation of the transmission. Who is correct?

    A. A only

    B. B only

    C. Both A and B

    D. Neither A nor B

40. Which of the following actions would be LEAST LIKELY to be performed prior to installing a torque converter?

    A. Add clean transmission fluid to the converter.

    B. Test the lockup clutch.

    C. Check the stator one-way roller clutch.

    D. Check the torque converter end-play.

41. All of the following actions are required in order to check the flow rate of the transmission fluid cooler assembly EXCEPT:

    A. Drain the engine coolant.

    B. Remove the cooler return line.

    C. Start the engine and run at 1,000 rpm.

    D. Measure the fluid flow for 20 seconds.

42. All of the following statements are correct concerning bearing preload EXCEPT:

    A. Bearing preload is sometimes checked by testing the turning effort of a component.

    B. Bearing preload that is set too tight will loosen up once the unit heats up.

    C. Bearing preload that is set loose will tighten up as the unit heats up.

    D. Bearings must be lubricated in order to perform under an extended load.

43. An encapsulated check ball has a worn sleeve. Technician A says that a larger check ball should be installed. Technician B says that the check ball and sleeve can be removed and replaced. Who is correct?

    A. A only

    B. B only

    C. Both A and B

    D. Neither A nor B

44. Technician A says that if a bushing shows wear it should be replaced. Technician B says that the shaft that mates with the bushing should also be checked if the bushing is worn. Who is correct?

    A. A only

    B. B only

    C. Both A and B

    D. Neither A nor B

45. A technician is diagnosing a transmission for a possible fractured case assembly. Technician A says that fluid can be poured into a suspect case bore to check for a leak. Technician B says that a fractured case can be repaired by welding the fractured area. Who is correct?

    A. A only

    B. B only

    C. Both A and B

    D. Neither A nor B

46. All of the following actions should be followed when installing a transmission oil pump into a transmission EXCEPT:

    A. The pump internal components should be lubricated with clean transmission fluid.

    B. The pump internal gear should measured with a micrometer.

    C. The pump fasteners should be tightened to the correct torque.

    D. The pump should be rotated after tightening the fasteners to check for tightness.

47. Technician A says that new friction discs should be soaked in engine oil prior to assembly of a clutch pack. Technician B says that the clutch pack clearance should be checked after building each clutch pack. Who is correct?

  A. A only

  B. B only

  C. Both A and B

  D. Neither A nor B

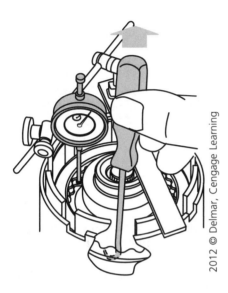

2012 © Delmar, Cengage Learning

48. Referring to the figure above, Technician A says that this test will measure clutch pack clearance. Technician B says that air pressure should be applied to the clutch pack during this test. Who is correct?

  A. A only

  B. B only

  C. Both A and B

  D. Neither A nor B

49. Technician A says that the inner race of a roller clutch should lock in one direction. Technician B says that the inner race of a roller clutch should freewheel in one direction. Who is correct?

  A. A only

  B. B only

  C. Both A and B

  D. Neither A nor B

50. What would be the most likely result of a transmission band that was adjusted too tightly?

  A. Vehicle will take off in the wrong gear.

  B. Vehicle will not move forward or reverse.

  C. Vehicle will burn a clutch pack from overslipping.

  D. Servo seal will leak externally.

PREPARATION EXAM 6

1. Which of the following conditions could cause the transmission to slip when the vehicle makes sharp turns?

   A. Brake switch is stuck open.

   B. Engine ignition system misfires.

   C. Transmission fluid is low.

   D. Transmission cooler is restricted.

2. A technician notices that all shifts are late and very firm during a road test. The vehicle has a hydraulic-shifted transmission and uses a vacuum modulator. Technician A says that the vacuum hose leading to the vacuum modulator could be plugged. Technician B says that there could be a leak in the vacuum hose leading to the vacuum modulator. Who is correct?

   A. A only

   B. B only

   C. Both A and B

   D. Neither A nor B

3. A technician notices that the engine exhaust system is emitting white smoke during a road test. All of the following conditions could cause this condition EXCEPT:

   A. Engine intake gasket is blown.

   B. Vacuum modulator is ruptured.

   C. Transmission cooler is plugged.

   D. Engine head gasket is blown.

4. Which of the following faults would be most likely to cause a clicking noise that increases when the transmission is shifted into drive and goes away when the transmission is shifted into park?

   A. Worn transmission pump

   B. Cracked flexplate

   C. Slipping transmission clutch pack

   D. Torque converter lockup clutch slipping

5. A vehicle is being diagnosed for a power train vibration at 50 to 60 mph. Technician A says that a slipping clutch pack could cause this problem. Technician B says that a faulty universal joint could cause this problem. Who is correct?

   A. A only

   B. B only

   C. Both A and B

   D. Neither A nor B

6. Which is the LEAST LIKELY factor that would be used to correctly check the transmission fluid?

   A. Vehicle on a level surface
   B. Engine at 1500 rpm
   C. Transmission in the correct gear position
   D. Transmission near operating temperature

7. A vehicle is being diagnosed that has extremely burned and discolored transmission fluid. Technician A says that burned clutch discs in a clutch pack could be the cause. Technician B says that a leak in the transmission cooler could be the cause. Who is correct?

   A. A only
   B. B only
   C. Both A and B
   D. Neither A nor B

8. A vehicle will back-up and go into first gear without a problem. The vehicle will not up-shift to second gear until very high engine RPM. Technician A says that the governor passage could be partially blocked. Technician B says that the second gear shift valve could be sticking. Who is correct?

   A. A only
   B. B only
   C. Both A and B
   D. Neither A nor B

9. Technician A says that a low stall speed could be caused by a slipping clutch pack assembly. Technician B says that a high stall speed could be caused by a slipping band. Who is correct?

   A. A only
   B. B only
   C. Both A and B
   D. Neither A nor B

10. Technician A says that the transmission computer controls the torque converter lockup clutch. Technician B says that an inoperative torque converter lockup clutch can cause a decrease in fuel economy. Who is correct?

    A. A only
    B. B only
    C. Both A and B
    D. Neither A nor B

11. An inoperative torque converter lockup clutch will cause all of the following results EXCEPT:

    A. Reduced fuel economy
    B. Increased engine temperature
    C. Increased engine RPM at highway speeds
    D. Diagnostic trouble code

12. Which of the following items should be monitored during the road test of a vehicle with an electronic transmission?

    A.   Engine oil temperature

    B.   Speed sensor resistance

    C.   Transmission up-shift points

    D.   Transmission fluid level

13. Technician A says that the governor pressure should be monitored during the road test of a vehicle with an electronic transmission. Technician B says that the vacuum should be checked at the modulator valve during the road test of a vehicle with an electronic transmission. Who is correct?

    A.   A only

    B.   B only

    C.   Both A and B

    D.   Neither A nor B

14. Technician A says the excessive line pressure can cause soft shifts. Technician B says that low line pressure can cause firm shifts. Who is correct?

    A.   A only

    B.   B only

    C.   Both A and B

    D.   Neither A nor B

15. A vehicle with an electronic transmission has been tested for transmission pump output pressure. The pressures attained during the test were above the specifications in all ranges. Which of the following conditions would be most likely the cause of this problem?

    A.   Restricted transmission filter

    B.   Loose terminal connection at the pressure control solenoid

    C.   Punctured transmission fluid pickup tube

    D.   Excess transmission pump clearance

16. A vehicle is being diagnosed for a problem of the engine dying when coming to a stop. After restarting the engine, it again dies when the transmission is shifted into any gear. Technician A says that the brake switch could be the cause. Technician B says that the torque converter clutch solenoid could be the cause. Who is correct?

    A.   A only

    B.   B only

    C.   Both A and B

    D.   Neither A nor B

17. All of the following conditions have to be met in order for the torque converter clutch to engage EXCEPT:

    A.   Light to moderate throttle application

    B.   Vehicle above the minimum converter clutch set speed

    C.   Transmission in second gear

    D.   Engine temperature minimum met

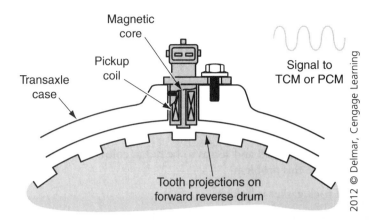

18. Referring to the figure above, all of the following statements are correct about the speed sensor above EXCEPT:

    A. The speed sensor contains a coil.

    B. The output frequency of the sensor increases as the forward/reverse drum slows down.

    C. The output voltage of the sensor increases as the forward/reverse drum speeds up.

    D. The speed sensor contains a permanent magnet.

19. Which of the following purposes would a wire schematic most likely be needed for when diagnosing an electronic transmission?

    A. Finding a flowchart for a trouble code

    B. Following the fluid flow in a hydraulic circuit

    C. Determining the pin identification for a connector on a transmission

    D. Finding the physical location of an electronic transmission component

20. Technician A says that a 12 volt battery that has 6 volts at the posts is 50 percent charged. Technician B says that a 12 volt battery that has 12.6 volts at the posts is overcharged. Who is correct?

    A. A only

    B. B only

    C. Both A and B

    D. Neither A nor B

21. A vehicle is being diagnosed with a transmission that stays in second gear all of the time. The supply (B+) voltage was tested at the transmission computer and was found to be at 10.5 volts with the engine running. Technician A says that a voltage drop test should be performed on the (B+) circuit. Technician B says that the problem could be caused by a wire repair made with wire that is too large. Who is correct?

    A. A only

    B. B only

    C. Both A and B

    D. Neither A nor B

A. Transmission solenoid pack gasket is leaking.

B. Engine intake gasket is leaking.

C. Transmission cooler is leaking.

D. Transmission modulator valve is leaking.

24. A vehicle with an electronic transmission will not move in forward or reverse. All of the following faults could cause this problem EXCEPT:

A. Broken input shaft

B. Faulty input shaft speed sensor

C. Low transmission fluid level

D. Faulty transmission oil pump

25. The main power fuse for the electronic control system has failed. Technician A says that the vehicle will still have one forward gear as well as reverse. Technician B says that this fault should set a diagnostic trouble code (DTC) in the transmission computer. Who is correct?

A. A only

B. B only

C. Both A and B

D. Neither A nor B

26. Which of the following components could cause the electronic shift indicator to show the incorrect gear?

A. Faulty park/neutral switch

B. Misadjusted range sensor

C. Misadjusted throttle position sensor

D. Faulty speed sensor

27. Technician A says that the bottom pan must be removed in order to replace the park/neutral switch. Technician B says that the range sensor can be replaced without removing the transmission pan. Who is correct?

A. A only

B. B only

C. Both A and B

D. Neither A nor B

28. The throttle cable needs to be replaced on a late-model vehicle. Technician A says that the kick-down cable may need to be adjusted after installing the new throttle cable. Technician B says that the throttle plate is typically spring loaded to the open position. Who is correct?

   A. A only
   B. B only
   C. Both A and B
   D. Neither A nor B

29. Which of the following statements best describes the action of the throttle valve linkage?

   A. The throttle valve linkage connects the engine to the cruise control servo.
   B. The throttle valve linkage connects the throttle pedal to the engine.
   C. The throttle valve provides vehicle speed feedback to the transmission valve body.
   D. The throttle valve provides engine load feedback to the transmission valve body.

30. Transmission fluid is leaking from the output speed sensor area. Technician A says that the speed sensor o-ring may be damaged. Technician B says that the speed sensor should be replaced. Who is correct?

   A. A only
   B. B only
   C. Both A and B
   D. Neither A nor B

31. Transmission fluid is leaking from the manual linkage near the transmission case. Technician A says that the transmission pan will need to be removed to replace the shift linkage seal. Technician B says that the shift linkage will need to be disassembled to replace the shift linkage seal. Who is correct?

   A. A only
   B. B only
   C. Both A and B
   D. Neither A nor B

32. A late-model vehicle is being diagnosed for a cooling system leak. Technician A says that applying pressure to the system with a pressure tester may be necessary. Technician B says that adding coolant dye to the system may be necessary. Who is correct?

   A. A only
   B. B only
   C. Both A and B
   D. Neither A nor B

33. Technician A says that the valve body mating surface should be cleaned with medium grit sandpaper. Technician B says that the valve body bores should be cleaned with a light shop rag. Who is correct?

   A. A only
   B. B only
   C. Both A and B
   D. Neither A nor B

A. The vehicle will never shift to second gear.
B. The vehicle will launch from a stop in second gear.
C. The vehicle will be stuck in reverse gear.
D. The vehicle will not pull in forward or reverse.

36. A vehicle is being diagnosed for a bumping noise that occurs when backing up a steep incline. What transmission component would most likely cause this problem?
    A. Chipped planetary gear
    B. Restricted transmission cooler
    C. Transmission band adjusted too loose
    D. Broken rear transmission mount

37. Where would the transmission filter be LEAST LIKELY located?
    A. Mounted in a transmission cooler line
    B. Mounted on the transmission case
    C. Mounted inside the transmission pan
    D. Mounted inside the torque converter

38. Technician says the front pump seal should be replaced any time that the torque converter is replaced. Technician B says that a new torque converter should be at least partly filled with new fluid prior to installing into the transmission. Who is correct?
    A. A only
    B. B only
    C. Both A and B
    D. Neither A nor B

39. Failing to completely install the torque converter into the transmission prior to installing the transmission will most likely cause which of the following results?
    A. Broken input shaft
    B. Front crankshaft seal to leak
    C. Damage to the clutch pack piston
    D. Damage to the transmission oil pump gears

40. Technician A says that the transmission cooler should be flushed every time that the transmission is rebuilt or replaced. Technician B says that the cooler should be flushed every 30,000 miles. Who is correct?
    A. A only
    B. B only
    C. Both A and B
    D. Neither A nor B

42. Technician A says that all bushings should be replaced during a transmission overhaul operation. Technician B says that most bushing material is softer than the shaft it rides against. Who is correct?

    A. A only
    B. B only
    C. Both A and B
    D. Neither A nor B

43. The planetary gear set is blue and discolored. Technician A says that the gear set has been overheated. Technician B says that the gear set should be replaced. Who is correct?

    A. A only
    B. B only
    C. Both A and B
    D. Neither A nor B

44. Technician A says that a transaxle drive chain should be replaced during every overhaul procedure. Technician B says that the chain should be greased heavily during assembly to prevent early chain failure. Who is correct?

    A. A only
    B. B only
    C. Both A and B
    D. Neither A nor B

45. Which of the following methods would most likely be used when testing the contact pattern on the final drive gears in a transaxle?

    A. Gear compound
    B. Slide caliper
    C. Outside micrometer
    D. Feeler gauge

46. Technician A says that clutch discs should be installed into the transmission dry. Technician B says that all bands should be installed dry. Who is correct?

    A. A only
    B. B only
    C. Both A and B
    D. Neither A nor B

47. All of the following actions should be made prior to the assembly of a clutch pack EXCEPT:

A. The friction discs should be soaked in clean transmission fluid.

B. The clutch pack apply plate should be machined on a lathe to polish the surface.

C. The clutch pack drum should be thoroughly cleaned and inspected.

D. The clutch pack piston seals should be installed and lubricated with assembly lube.

2012 © Delmar, Cengage Learning

48. The test being performed in the figure above will provide feedback to the technician about all of the following clutch pack components EXCEPT:

A. Clutch pack piston

B. Clutch pack piston seal

C. Piston return spring

D. Clutch pack apply plate

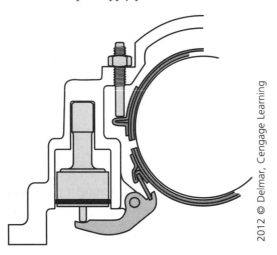

2012 © Delmar, Cengage Learning

49. All of the following components are present in the figure above EXCEPT:

A. Transmission band adjusting bolt and nut

B. Transmission band

C. Valve body

D. Transmission servo

50. What would be the most likely result of a transmission band that was adjusted too loosely?

A. Vehicle will take off in the wrong gear.

B. Vehicle will not move forward or reverse.

C. The transmission will have a slip or flare when the band is applied.

D. The servo seal will leak externally.

Answer Keys and Explanations

INTRODUCTION

Included in this section are the answer keys for each preparation exam, followed by individual, detailed answer explanations and a reference identifying the designated task area being assessed by each specific question. This additional reference information may prove useful if you need to refer back to the task list located in Section 4 of this book for additional support.

PREPARATION EXAM 1—ANSWER KEY

1.	C	21.	C	41.	C
2.	A	22.	B	42.	D
3.	C	23.	A	43.	D
4.	C	24.	D	44.	C
5.	A	25.	C	45.	A
6.	B	26.	B	46.	B
7.	C	27.	B	47.	A
8.	D	28.	B	48.	A
9.	B	29.	D	49.	C
10.	D	30.	D	50.	C
11.	B	31.	C		
12.	A	32.	C		
13.	B	33.	C		
14.	D	34.	D		
15.	C	35.	C		
16.	C	36.	D		
17.	C	37.	C		
18.	B	38.	B		
19.	C	39.	C		
20.	A	40.	C		

PREPARATION EXAM 1—EXPLANATIONS

TASK A.1.1

1. A late-model vehicle needs to be road tested to verify a shifting concern. Technician A says that the customer complaint should be read carefully prior to beginning the road test. Technician B says that the transmission fluid should be checked prior to beginning the road test. Who is correct?

    A.   A only
    B.   B only
    C.   Both A and B
    D.   Neither A nor B

    Answer A is incorrect. Technician B is also correct.

    Answer B is incorrect. Technician A is also correct.

    Answer C is correct. Both Technicians are correct. It is advisable to closely read the exact details of the customer complaint prior to beginning the road test in order to put the vehicle in the same conditions that the customer is putting it in. In addition, inspecting the transmission fluid will provide some needed information about the state of the transmission. Checking the fluid will also prevent the technician from driving the vehicle with low fluid, which could cause more damage.

    Answer D is incorrect. Both Technicians are correct.

TASK A.1.1

2. Which of the following faults would be most likely to cause a shudder condition during the road test?

    A.   Torque converter lockup clutch
    B.   Weak transmission oil pump
    C.   Planetary gearset chipped
    D.   Torque converter stator slipping

    Answer A is correct. A problem with the torque converter lockup clutch can cause a shudder condition. This problem will usually happen as the torque converter clutch begins to engage.

    Answer B is incorrect. A weak transmission oil pump would cause low transmission pressures and would likely cause the clutches to slip.

    Answer C is incorrect. A chipped planetary gear would possibly cause a noise while driving, but it would not cause a shudder condition.

    Answer D is incorrect. A slipping torque converter stator would cause low power at low vehicle speeds.

Answer A is incorrect. Technician B is also correct.

Answer B is incorrect. Technician A is also correct.

Answer C is correct. Both Technicians are correct. A problem in the constant velocity joint will typically be most noticeable while making turns. A faulty wheel hub bearing will make a roaring sound while making turns in the vehicle.

Answer D is incorrect. Both Technicians are correct.

4. Which of the following faults would most likely cause a vibration at 50 to 60 mph?

    A.   Low line pressure

    B.   Slipping clutch pack

    C.   Universal joint

    D.   Output shaft seal

TASK A.1.2

Answer A is incorrect. Low line pressure would cause the transmission to slip and not pull.

Answer B is incorrect. A slipping clutch pack would cause engine RPM to increase without a change in pulling torque.

Answer C is correct. A bad universal joint would cause a major vibration at highway speeds. This vibration can get very violent and is usually felt in the floor board and even in the driver's seat.

Answer D is incorrect. An output shaft seal problem would cause a leak at the rear of the transmission.

5. A vehicle is being diagnosed for a transmission fluid leak near the bell housing area. Technician A says that a failed front pump seal could cause a leak in this area. Technician B says that a leaking transmission cooler could cause a leak in this area. Who is correct?

    A.   A only

    B.   B only

    C.   Both A and B

    D.   Neither A nor B

TASK A.1.3

Answer A is correct. Only Technician A is correct. A failed front pump seal will cause transmission fluid to be present in the bell housing area.

Answer B is incorrect. A leaking transmission cooler will cause the transmission fluid to mix with the engine coolant.

Answer C is incorrect. Only Technician A is correct.

Answer D is incorrect. Technician A is correct.

TASK A.1.3

6. All of the following problems could cause a low transmission fluid level EXCEPT:

   A.   Ruptured transmission cooler
   B.   Slipping clutch pack
   C.   Faulty modulator valve
   D.   Front pump seal

   Answer A is incorrect. A ruptured transmission cooler can cause transmission fluid to be pumped into the cooling system.

   Answer B is correct. A slipping clutch pack will cause the engine speed to increase without any torque being transferred, but it will not cause a loss of transmission fluid.

   Answer C is incorrect. A modulator sometimes fails by letting transmission fluid be sucked into the engine vacuum hose.

   Answer D is incorrect. A front pump seal can fail by leaking transmission fluid into the bell housing area.

TASK A.1.4

7. Which of the following conditions could cause a vehicle to have reverse and first gears but not any up-shifts?

   A.   Low line pressure
   B.   Excessive line pressure
   C.   Low governor valve pressure
   D.   Excessive servo pressure

   Answer A is incorrect. Low line pressure will cause the transmission to either fail to pull at all or possibly cause a major slipping problem.

   Answer B is incorrect. Excessive line pressure would likely cause harsh shifts and possibly a noticeable transmission pump noise. The transmission would still be able to up-shift.

   Answer C is correct. A governor that is not creating enough pressure will cause the vehicle to never up-shift.

   Answer D is incorrect. Excessive servo pressure would likely not cause a noticeable problem.

TASK A.1.4

8. Which gear should a technician put the transmission in to create maximum line pressure?

   A.   Manual second
   B.   Drive
   C.   Park
   D.   Reverse

   Answer A is incorrect. Line pressure would not typically be elevated above normal in manual second gear.

   Answer B is incorrect. Line pressure would not be elevated in drive gear.

   Answer C is incorrect. Line pressure would likely be at the minimum pressure when the transmission is in park.

   Answer D is correct. Line pressure is typically at its highest level when the transmission is in reverse.

Answer A is incorrect. A slipping clutch pack would likely cause the stall speed to be higher than the specification.

Answer B is correct. Only Technician B is correct. A plugged exhaust system will cause low engine power as well as lower than normal stall speeds.

Answer C is incorrect. Only Technician B is correct.

Answer D is incorrect. Technician B is correct.

10. Technician A says that a stall test should be performed before the engine warms up. Technician B says performing a stall test for a minimum of two consecutive minutes is recommended. Who is correct?

TASK A.1.5

 A. A only
 B. B only
 C. Both A and B
 D. Neither A nor B

Answer A is incorrect. The engine should be at normal operating temperature before performing a stall test.

Answer B is incorrect. A stall test should only be performed for a few seconds at a time. Running this test for extended time will cause the transmission fluid to overheat.

Answer C is incorrect. Neither Technician is correct.

Answer D is correct. Neither Technician is correct. A stall test should be performed on a warmed-up vehicle and only for a few seconds at a time.

11. Technician A says that the engine RPM should drop 600 rpm when the torque converter clutch engages. Technician B says that the torque converter should disengage when the brake pedal is depressed. Who is correct?

TASK A.1.6

 A. A only
 B. B only
 C. Both A and B
 D. Neither A nor B

Answer A is incorrect. The RPM will only drop about 150 to 200 rpm when the torque converter clutch engages.

Answer B is correct. Only Technician B is correct. Applying the brake pedal will typically cause the torque converter clutch to disengage. Torque converter clutch systems are designed this way in order to prevent the engine from dying when coming to a stop.

Answer C is incorrect. Only Technician B is correct.

Answer D is incorrect. Technician B is correct.

12. A vehicle with an electronic shifted transmission is being road tested. Technician A says that a scan tool can be useful to use during the road test. Technician B says that a pressure manifold can be useful during the road test. Who is correct?

    A. A only

    B. B only

    C. Both A and B

    D. Neither A nor B

    Answer A is correct. Only Technician A is correct. A scan tool is useful when road testing a vehicle with an electronic shifted transmission. The scanner can be used to retrieve trouble codes and data from the transmission computer.

    Answer B is incorrect. It is not common to take a pressure manifold on a road test for a transmission complaint.

    Answer C is incorrect. Only Technician A is correct.

    Answer D is incorrect. Technician A is correct.

13. Which of the following activities would be LEAST LIKELY to be performed during the road test of a vehicle with an electronic transmission?

    A. Monitor the shift points of all forward gears.

    B. Monitor the vacuum level to the modulator.

    C. Listen for unusual noises in the drive train.

    D. Observe the feel of the downshift pattern.

    Answer A is incorrect. It is advisable to pay attention to the shift points of each shift during the road test.

    Answer B is correct. It would not be common to monitor any vacuum levels during a road test. In addition, most electronic transmissions would not even use a vacuum modulator.

    Answer C is incorrect. It is advisable to listen to the sound of the engine, transmission, and drive train components during the road test.

    Answer D is incorrect. It is advisable to pay attention to the downshift pattern and feel during the road test.

14. A pressure control solenoid has failed. This may cause all of the following EXCEPT:

    A. A DTC will set.

    B. The main line pressure will increase.

    C. The transmission will shift hard.

    D. The transmission will have no forward gears.

    Answer A is incorrect. A diagnostic trouble code (DTC) will usually set when the pressure control solenoid fails.

    Answer B is incorrect. The pressure control solenoid is typically designed to cause high transmission pressure if the solenoid fails. This strategy is a fail-safe design that most manufacturers utilize.

    Answer C is incorrect. The transmission will likely shift hard when the pressure control solenoid fails. This harsh shifting will be a result of high line pressure.

    Answer D is correct. A failed pressure control solenoid will not cause the transmission to lose its ability to pull. The line pressure will typically be increased when the solenoid fails.

D.   Neither A nor B

Answer A is incorrect. Technician B is also correct.

Answer B is incorrect. Technician A is also correct.

Answer C is correct. Both Technicians are correct. A disconnected pressure control solenoid will cause the line pressure to increase to the maximum level. This high line pressure will cause all garage shifts to be sharp and jerky.

Answer D is incorrect. Both Technicians are correct.

16. All of the following conditions have to be met in order for the torque converter clutch to engage EXCEPT:

TASK A.2.3

A.   Brakes are not applied.
B.   Vehicle speed minimum is met.
C.   Transmission is in first gear.
D.   Engine temperature minimum is met.

Answer A is incorrect. The torque converter will not engage with the brakes applied.

Answer B is incorrect. The torque converter will not engage until a minimum speed has been reached. This minimum speed is usually around 35 miles per hour.

Answer C is correct. The torque converter clutch will never be engaged in first gear. Typically, the torque converter clutch will only engage in the top two gears.

Answer D is incorrect. The torque converter clutch will not engage when the engine is below a certain temperature.

17. A vehicle is being diagnosed for a drop in fuel economy. Technician A says that an inoperative torque converter clutch could be the cause. Technician B says that a stuck open engine thermostat could be the cause. Who is correct?

TASK A.2.3

A.   A only
B.   B only
C.   Both A and B
D.   Neither A nor B

Answer A is incorrect. Technician B is also correct.

Answer B is incorrect. Technician A is also correct.

Answer C is correct. Both Technicians are correct. Vehicle fuel economy will suffer if the torque converter clutch is inoperative due to the extra engine RPMs that the engine will have to endure. A stuck open engine thermostat will also cause a drop in fuel economy due to the extended warm-up time that would result here.

Answer D is incorrect. Both Technicians are correct.

TASK A.2.4

18. A transmission fluid temperature (TFT) sensor is being tested. Technician A says that an ammeter should be used to test this device. Technician B says that the resistance of the sensor should decrease as the sensor is heated. Who is correct?

   A. A only
   B. B only
   C. Both A and B
   D. Neither A nor B

   Answer A is incorrect. An ohmmeter is the tool used to test a transmission fluid temperature sensor.

   Answer B is correct. Only Technician B is correct. Transmission fluid temperature sensors are usually negative temperature coefficient (NTC) thermistors. These devices drop in resistance as the temperature increases.

   Answer C is incorrect. Only Technician B is correct.

   Answer D is incorrect. Technician B is correct.

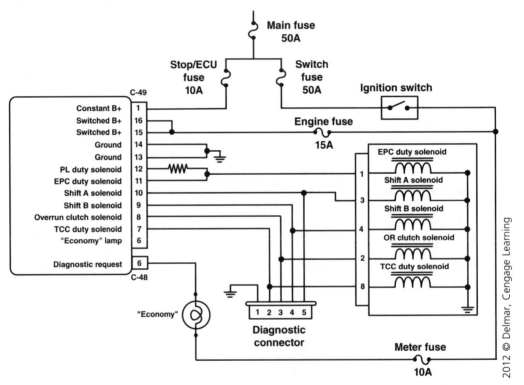

19. Referring to the figure above, Technician A says that all of the solenoids receive ground from the transmission case. Technician B says that the transmission EPC duty solenoid receives power from pins 11 and 12 of the transmission computer. Who is correct?

TASK A.2.4

   A. A only
   B. B only
   C. Both A and B
   D. Neither A nor B

   Answer A is incorrect. Technician B is also correct.

   Answer B is incorrect. Technician A is also correct.

   Answer C is correct. Both Technicians are correct. The solenoids in the schematic are case grounded. The pressure control solenoid does receive power from pins 11 and 12 of the transmission computer.

   Answer D is incorrect. Both Technicians are correct.

D. Neither A nor B

Answer A is correct. Only Technician A is correct. The replacement generator should be identical to the unit being replaced. The Technician should make sure that the pulley is the same diameter as well as have the same number of grooves for the belt.

Answer B is incorrect. It is not necessary to disassemble and test the replacement generator before installing on the vehicle.

Answer C is incorrect. Only Technician A is correct.

Answer D is incorrect. Technician A is correct.

21. Which of the following conditions would be the most likely condition to cause the generator to overcharge?

TASK A.2.5

A. Blown fusible link in charge circuit

B. An open rotor coil

C. A full-fielded rotor

D. A faulty voltage regulator

Answer A is incorrect. A blown fusible link in the charge circuit would cause the generator to not charge at all.

Answer B is incorrect. An open rotor coil would cause the generator to not charge at all.

Answer C is correct. A full-fielded rotor would cause the generator to overcharge at all times.

Answer D is incorrect. A faulty voltage regulator could cause the generator to overcharge, undercharge, or not charge at all.

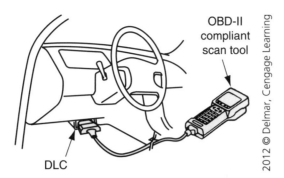

OBD-II compliant scan tool

2012 © Delmar, Cengage Learning

DLC

TASK A.2.6

22. The tool shown in the picture above can be used for which of the following purposes?

A. Checking for the supply voltage (B+) at the transmission computer

B. Accessing the diagnostic trouble codes (DTCs) in the transmission computer

C. Testing the resistance of the speeds sensor coil

D. Testing the voltage of the speeds sensor

Answer A is incorrect. A digital voltmeter would be needed to check for supply voltage at the transmission computer.

Answer B is correct. The scan tool in the picture can be used to access diagnostic trouble codes (DTCs) in the transmission computer. In addition, the scan tool can retrieve data, as well as command output tests on the solenoids and indicators.

Answer C is incorrect. A digital ohmmeter would be needed to test the speed sensor resistance.

Answer D is incorrect. A digital voltmeter would be needed to test the output voltage of the speed sensor.

TASK A.2.6

23. A vehicle is being diagnosed for a jerking condition that is most prevalent in overdrive. Technician A says that a faulty spark plug could be the cause. Technician B says that a slipping clutch pack in the transmission could be the cause. Who is correct?

A. A only

B. B only

C. Both A and B

D. Neither A nor B

Answer A is correct. Only Technician A is correct. A faulty spark plug will cause the vehicle to jerk very badly in overdrive. Engine misfires are most noticeable in the transmission's highest gear due to the increased load on the engine.

Answer B is incorrect. A slipping transmission clutch pack would cause the engine RPM to increase without adding output torque.

Answer C is incorrect. Only Technician A is correct.

Answer D is incorrect. Technician A is correct.

24. Referring to the figure above, Technician A says the pattern on Channel 1 of the oscilloscope is a good signal from a permanent magnet generator (speed sensor). Technician B says the pattern from Channel 2 is a good signal from a "hall effect" sensor. Who is correct?

    A.   A only

    B.   B only

    C.   Both A and B

    D.   Neither A nor B

TASK A.2.7

Answer A is incorrect. Channel 1 of this oscilloscope shows a digital (square wave) signal. A "hall effect" sensor often has this type of signal.

Answer B is incorrect. Channel 2 of this oscilloscope shows an analog signal. This analog signal is typical for a permanent magnet generator.

Answer C is incorrect. Neither Technician is correct.

Answer D is correct. Neither Technician is correct. Hall effect sensors produce a digital (square wave) signal like the example on Channel 1. Permanent magnet generators produce a sine wave like the example on channel 2.

25. Technician A says that a scan tool can be used to retrieve trouble codes from the transmission computer. Technician B says that a scan tool can be used to retrieve sensor data from the transmission computer. Who is correct?

    A.   A only

    B.   B only

    C.   Both A and B

    D.   Neither A nor B

TASK A.2.7

Answer A is incorrect. Technician B is also correct.

Answer B is incorrect. Technician A is also correct.

Answer C is correct. Both Technicians are correct. Scan tools will retrieve data as well as trouble codes from the transmission computer. Scan tools can also perform output tests on many of the output components connected to the computer.

Answer D is incorrect. Both Technicians are correct.

TASK B.1

26. A vehicle is being diagnosed for a problem of the transmission not engaging into manual low gear. The transmission will engage into all other gears without a problem. Technician A says that the shift indicator could be misaligned. Technician B says that the shift linkage could be binding. Who is correct?

A.   A only
B.   B only
C.   Both A and B
D.   Neither A nor B

Answer A is incorrect. A shift indicator would not prevent the driver from moving the gear shift into manual low gear.

Answer B is correct. Only Technician B is correct. A binding shift linkage could restrict the manual valve from moving into the manual low position.

Answer C is incorrect. Only Technician B is correct.

Answer D is incorrect. Technician B is correct.

TASK B.2

27. A transaxle is being diagnosed for a late shifting problem. Which of the following faults would most likely cause this problem?

A.   Broken throttle valve cable
B.   Throttle valve linkage adjusted too tight
C.   Loose band adjustment
D.   Throttle valve linkage adjusted too loose

Answer A is incorrect. A broken throttle valve cable would cause the transmission up-shifts to be very early.

Answer B is correct. A tight throttle valve linkage adjustment could cause the up-shifts to occur later than normal.

Answer C is incorrect. A loose band adjustment would cause slip or a flare, but it would not cause a late shifting problem.

Answer D is incorrect. Loose throttle valve linkage adjustments would cause early up-shifts.

TASK B.3

28. A transmission is leaking near the area where the drive shaft enters the transmission. Which of the components below is the most likely cause of this leak?

A.   Oil pump seal
B.   Output shaft seal
C.   Manual linkage seal
D.   Input speed sensor seal

Answer A is incorrect. A leaking oil pump seal would cause fluid to leak around the bell housing area.

Answer B is correct. A leaking output shaft seal will cause a leak at the output shaft area.

Answer C is incorrect. A leaking manual valve linkage seal would cause a fluid leak on the side of the transmission.

Answer D is incorrect. A leaking input speed sensor seal would cause a leak near the front of the transmission.

Answer A is incorrect. A worn differential drive pinion would cause a roaring sound while the vehicle was being driven.

Answer B is incorrect. An overfilled transmission could cause aeration of the fluid which could cause the transmission to slip.

Answer C is incorrect. Worn u-joints at the differential could cause pinion seal failure.

Answer D is correct. A worn output shaft support bushing would cause the output shaft to have excessive axial movement which would cause heavy wear on the output shaft seal.

30. The windshield fogs up when the defroster is turned on and the cab is filled with a sweet smell. Which of the following is the most likely cause?

    A.   Blown head gasket

    B.   Leaking heater control valve

    C.   Leaking evaporator core

    D.   Leaking heater core

TASK B.5

Answer A is incorrect. A blown head gasket on the engine would not cause the fogging problem or the sweet smell in the cab.

Answer B is incorrect. A leaking heater control valve would not cause any noticeable problems inside the cab.

Answer C is incorrect. A leaking evaporator core would not produce a sweet smell in the cab.

Answer D is correct. A leaking heater core can cause windshield fogging as well as sometimes producing a sweet smell from the coolant.

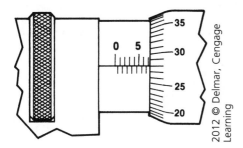

2012 © Delmar, Cengage Learning

31. What is the measurement reading on the metric micrometer above?

    A.   9.28 mm

    B.   7.78 mm

    C.   7.28 mm

    D.   9.78 mm

TASK B.6

Answer A is incorrect. The metric micrometer measurement in the picture is 7.28 mm.

Answer B is incorrect. The metric micrometer measurement in the picture is 7.28 mm.

Answer C is correct. The metric micrometer measurement in the picture is 7.28 mm.

Answer D is incorrect. The metric micrometer measurement in the picture is 7.28 mm.

TASK B.7

32. Which of the following practices would be LEAST LIKELY to be followed when installing the bolts into a valve body?

    A.   Start all bolts by hand.

    B.   Tighten all bolts gradually.

    C.   Apply thread locker to all valve body bolt threads.

    D.   Use the specified torque wrench to complete the tightening sequence.

    Answer A is incorrect. All valve body bolts should be started by hand in order to prevent damage to the threads from cross threading.

    Answer B is incorrect. It is advisable to gradually tighten valve body bolts in order to secure the valve body in a uniform manner.

    Answer C is correct. Thread locker should only be used when the transmission manufacturer recommends it. Most valve body bolts should be assembled clean and dry.

    Answer D is incorrect. The specified torque wrench should always be used in the last stage of the tightening procedure.

TASK B.8

33. Technician A says that a servo can be air tested to test for proper piston sealing. Technician B says that the servo spring causes the servo to release when pressure is released from the piston. Who is correct?

    A.   A only

    B.   B only

    C.   Both A and B

    D.   Neither A nor B

    Answer A is incorrect. Technician B is also correct.

    Answer B is incorrect. Technician A is also correct.

    Answer C is correct. Both Technicians are correct. Servo assemblies can be air tested to determine if the piston seal is functioning correctly. The servo spring acts to release the piston when pressure is released from the piston.

    Answer D is incorrect. Both Technicians are correct.

TASK B.9

34. Which of the following details would be the most likely item to be located on a wiring diagram for an electronic transmission?

    A.   The location of a component

    B.   A flowchart for troubleshooting an electrical problem

    C.   The current rating of a circuit

    D.   The color and circuit number of a wire

    Answer A is incorrect. A wiring diagram does not typically give the location of the electrical components.

    Answer B is incorrect. A wiring diagram does not typically provide a troubleshooting flowchart. However, an experienced technician can use the wiring diagram to develop a strategy for solving electrical problems.

    Answer C is incorrect. A wiring diagram does not typically provide information about the current rating of the circuit. However, an experienced technician can observe the circuit protection devices in the diagram and recognize the limits that cause these devices to open the circuit.

    Answer D is correct. Wiring diagrams usually provide the color of each wire, as well as the circuit identification number of each wire.

Answer B is incorrect. A terminal is defined as a metallic component that is used to tie electrical circuits together

Answer C is correct. A connector is a plastic housing that holds terminals, as well as plugs, in electrical items such as motors and modules. Connectors often plug into other connectors.

Answer D is incorrect. A splice is defined as a joint where two or more electrical circuits connect to each other.

36. What would most likely cause the engine to twist excessively when accelerating quickly?

   A.   Loose transmission bell housing bolts
   B.   Broken transmission mount
   C.   Loose engine accessory bracket
   D.   Broken engine mount

TASK B.10

Answer A is incorrect. A loose transmission bell housing bolt could cause the transmission to become loose, which could cause a noise or even a fluid leak.

Answer B is incorrect. A broken transmission mount could cause the transmission to move excessively when accelerating. In addition, a broken transmission mount can cause a knocking or bumping sound under the vehicle.

Answer C is incorrect. A loose engine accessory bracket could cause noises and possibly excessive drive belt wear.

Answer D is correct. A broken engine mount could cause the engine to twist when the vehicle is quickly accelerated.

37. Which gear is most likely to be selected when checking the transmission fluid level?

   A.   Manual low
   B.   Reverse
   C.   Neutral
   D.   Overdrive

TASK B.11

Answer A is incorrect. It is not normal to put the transmission into manual low when checking the transmission fluid level.

Answer B is incorrect. It is not normal to put the transmission into reverse when checking the transmission fluid level.

Answer C is correct. Some manufacturers require the transmission be put into neutral when checking the transmission fluid.

Answer D is incorrect. It is not normal to put the transmission into overdrive when checking the transmission fluid.

TASK C.1.1

38. Which of the following components would be LEAST LIKELY to be removed during a transmission removal process?

A.   Torque converter bolts

B.   Alternator

C.   Bell housing bolts

D.   Negative battery cable

Answer A is incorrect. The torque converter bolts would be removed during a transmission removal in order to disconnect the converter from the flex plate.

Answer B is correct. The alternator would not typically need to be removed during a transmission removal.

Answer C is incorrect. The bell housing bolts would be removed during a transmission removal in order to disconnect the engine block from the transmission case.

Answer D is incorrect. The negative battery cable would be removed during a transmission removal in order to keep from shorting out the positive battery cable when working around the starter.

TASK C.1.2

39. Technician A says that the transmission needs to be removed to closely inspect the converter flexplate. Technician B says that the starter ring gear should be closely inspected for tooth damage when servicing the flexplate. Who is correct?

A.   A only

B.   B only

C.   Both A and B

D.   Neither A nor B

Answer A is incorrect. Technician B is also correct.

Answer B is incorrect. Technician A is also correct.

Answer C is correct. Both Technicians are correct. It would be necessary to remove the transmission in order to closely inspect the converter flexplate. The starter ring gear should always be closely inspected when servicing the flexplate.

Answer D is incorrect. Both Technicians are correct.

TASK C.1.2

40. Technician A says that a torque converter that has excessive end-play should be replaced. Technician B says that a torque with a failed one-way clutch should be replaced. Who is correct?

A.   A only

B.   B only

C.   Both A and B

D.   Neither A nor B

Answer A is incorrect. Technician B is also correct.

Answer B is incorrect. Technician A is also correct.

Answer C is correct. Both Technicians are correct. Excessive end-play or a failed one-way clutch is a necessary reason to replace the torque converter.

Answer D is incorrect. Both Technicians are correct.

Outlet
connector

2012 © Delmar, Cengage Learning

41. Referring to the figure above, Technician A says the operation in the figure should be performed with the engine at approximately 1,000 rpm. Technician B says that if the volume for this test is less than one quart of fluid in 20 seconds, the transmission cooler should be flushed. Who is correct?

TASK C.1.3

A. A only

B. B only

C. Both A and B

D. Neither A nor B

Answer A is incorrect. Technician B is also correct.

Answer B is incorrect. Technician A is also correct.

Answer C is correct. Both Technicians are correct. The transmission cooler flow test should be performed at approximately 1,000 rpm and should produce approximately one quart in 20 seconds. This test should be performed when there is evidence that foreign materials could have entered the cooler.

Answer D is incorrect. Both Technicians are correct.

42. Which of the following methods would be LEAST LIKELY used in cleaning the mating surface of a valve body and its mating transmission case surface?

TASK C.2.1

A. Compressed air

B. Aerosol-based solution

C. Solvent-based solution

D. Shop rag

Answer A is incorrect. Compressed air could be used to clean a valve body mating surfaces and oil passages without causing any problems.

Answer B is incorrect. An aerosol-based solution could be used to clean a valve body mating surface without causing any problems.

Answer C is incorrect. A solvent-based solution could be used to clean a valve body mating surface without causing any problems.

Answer D is correct. A shop rag should never be used to clean any of the valve body parts. A piece of lint from the shop rag could get stuck in the small passages of the valve body.

TASK C.2.2

43. Technician A says that an oil pump cover can be checked for flatness with a dial indicator. Technician B says that oil pump gear-to-gear clearance can be checked with a straightedge. Who is correct?

  A. A only
  B. B only
  C. Both A and B
  D. Neither A nor B

Answer A is incorrect. A straightedge and feeler gauge would be needed to check the flatness of an oil pump cover.

Answer B is incorrect. Oil pump gear-to-gear clearance can be checked with a feeler gauge.

Answer C is incorrect. Neither Technician is correct.

Answer D is correct. Neither Technician is correct. A straightedge and feeler gauge would be needed to check the flatness of an oil pump cover. A feeler gauge would be used to check the clearance on a gear-style oil pump.

TASK C.2.4

44. Technician A says that thrust washers should be measured with a dial caliper to check for correct thickness. Technician B says that some thrust washers are selective and are used to change the end-play on some transmission shafts. Who is correct?

  A. A only
  B. B only
  C. Both A and B
  D. Neither A nor B

Answer A is incorrect. Technician B is also correct.

Answer B is incorrect. Technician A is also correct.

Answer C is correct. Both Technicians are correct. Thrust washers should be measured with a precision measuring instrument, such as a dial caliper. Some thrust washers come in different thicknesses in order to change the end-play on the transmission.

Answer D is incorrect. Both Technicians are correct.

TASK C.2.6

45. A technician suspects a piece of debris is lodged in an oil delivery circuit in a transmission case. Which of the following methods would be most likely successful in clearing the debris from this area?

  A. Solvent and compressed air
  B. Pressurized nitrogen
  C. Water
  D. Shop rag and a piece of wire

Answer A is correct. Using solvent and compressed air would likely dislodge a piece of debris from an oil delivery circuit.

Answer B is incorrect. It is not common to use pressurized nitrogen on a transmission case.

Answer C is incorrect. It is not common to use water to clean a transmission case.

Answer D is incorrect. A shop rag should never be used on a transmission oil circuit. A piece of lint from the rag could get easily lodged in the circuit.

D.   Neither A nor B

Answer A is incorrect. The new torque converter should most always rest at the same location as the old torque converter. Care should be taken to correctly install the torque converter all the way into the transmission.

Answer B is correct. Only Technician B is correct. If the torque converter does not fully engage into the transmission, the technician should continue rotating and pushing the converter until it fully seats.

Answer C is incorrect. Only Technician B is correct.

Answer D is incorrect. Technician B is correct.

47. Technician A says clutch pack clearance can be decreased by adding a thicker shim (reaction/pressure plate). Technician B says clutch pack clearance can be decreased by installing a thicker apply piston. Who is correct?

TASK C.3.2

A.   A only

B.   B only

C.   Both A and B

D.   Neither A nor B

Answer A is correct. Only Technician A is correct. A thicker reaction/pressure plate can be installed to tighten the clearance in a clutch pack. The extra thickness of the plate would decrease the clearance in the clutch pack.

Answer B is incorrect. Clutch pack clearance is not adjusted by changing to a different size apply piston in the clutch pack.

Answer C is incorrect. Only Technician A is correct.

Answer D is incorrect. Technician A is correct.

48. Which of the following faults would be LEAST LIKELY to be discovered during a servo air test?

TASK C.3.3

A.   Faulty servo apply pin length

B.   Cracked transmission passage

C.   Servo piston

D.   Servo piston seal

Answer A is correct. Performing a servo air test would not likely reveal a faulty apply pin length.

Answer B is incorrect. It would be possible to find a cracked transmission passage during a servo air test.

Answer C is incorrect. A fault in the servo piston could be revealed during a servo air test. There would be a large air leak if the piston were damaged.

Answer D is incorrect. A servo piston seal could be revealed during a servo air test by allowing air to escape around the seal.

TASK C.3.4

49. Technician A says that the inner race of a sprag clutch should lock in one direction. Technician B says that the inner race of a sprag clutch should freewheel in one direction. Who is correct?

   A.   A only
   B.   B only
   C.   Both A and B
   D.   Neither A nor B

Answer A is incorrect. Technician B is also correct.

Answer B is incorrect. Technician A is also correct.

Answer C is correct. Both Technicians are correct. Sprag clutches are used as holding devices in automatic transmissions. The correct function of a sprag clutch includes the inner race locking in one direction and freewheeling in the other direction.

Answer D is incorrect. Both Technicians are correct.

TASK C.3.5

50. Technician A says a transmission drum can be damaged by a slipping band. Technician B says that a new band should be soaked in clean transmission fluid before installation. Who is correct?

   A.   A only
   B.   B only
   C.   Both A and B
   D.   Neither A nor B

Answer A is incorrect. Technician B is also correct.

Answer B is incorrect. Technician A is also correct.

Answer C is correct. Both Technicians are correct. A slipping band will quickly damage the drum surface that it rides against. All friction devices such as bands and clutch discs should be soaked with clean transmission fluid prior to installation.

Answer D is incorrect. Both Technicians are correct.

5.	B	25.	C	45.	C
6.	C	26.	C	46.	C
7.	C	27.	A	47.	C
8.	D	28.	B	48.	C
9.	C	29.	B	49.	B
10.	C	30.	A	50.	B
11.	C	31.	B		
12.	B	32.	A		
13.	A	33.	A		
14.	C	34.	B		
15.	B	35.	B		
16.	C	36.	C		
17.	C	37.	B		
18.	B	38.	D		
19.	B	39.	A		
20.	A	40.	B		

PREPARATION EXAM 2—EXPLANATIONS

1. A late-model vehicle needs to be road tested to verify a shifting concern. Technician A says that the shift linkage should be adjusted prior to beginning the road test. Technician B says that the transmission fluid should be checked prior to beginning the road test. Who is correct?

TASK A.1.1

   A.   A only
   B.   B only
   C.   Both A and B
   D.   Neither A nor B

Answer A is incorrect. It would not be a common practice to perform any adjustments prior to performing the initial road test.

Answer B is correct. Only Technician B is correct. It is advisable to check the transmission fluid prior to beginning the road test in order to make sure that the level is sufficient to safely drive the vehicle. In addition, a technician can get a great amount of diagnostic data from inspecting the transmission fluid.

Answer C is incorrect. Only Technician B is correct.

Answer D is incorrect. Technician B is correct.

2. A damaged transmission mount could cause all of the following problems EXCEPT:

   A. Transmission not shifting into overdrive
   B. Knocking sound when the vehicle is shifted into drive
   C. Knocking sound when the vehicle is shifted into reverse
   D. Vibration during heavy throttle events

   Answer A is correct. A damaged transmission would not prevent the transmission from shifting into overdrive.

   Answer B is incorrect. A damaged transmission mount could sometimes cause a knocking sound when the vehicle is shifted into drive because of the extra movement allowed by the mount.

   Answer C is incorrect. A damaged transmission mount could sometimes cause a knocking sound when the vehicle is shifted into reverse because of the extra movement allowed by the mount.

   Answer D is incorrect. A damaged transmission mount could cause a vibration during heavy throttle events due to the extra movement allowed by the mount.

3. Which of the following faults would most likely cause a vibration at 40 to 45 mph while turning the steering wheel?

   A. Low line pressure
   B. Slipping clutch pack
   C. Constant velocity (CV) joint
   D. Output shaft seal

   Answer A is incorrect. Low transmission line pressure can cause a transmission to slip or even fail to pull at all.

   Answer B is incorrect. A slipping clutch pack will cause the engine RPM to increase without a transfer of torque.

   Answer C is correct. A bad CV joint will typically cause a vibration while making a turn.

   Answer D is incorrect. A faulty output shaft seal will cause the transmission fluid to leak from the rear of the transmission.

4. Technician A says that a faulty clutch pack piston seal can cause a clicking noise when the vehicle is making a turn. Technician B says that a faulty torque converter stator roller clutch can cause a metallic roar while making a turn. Who is correct?

   A. A only
   B. B only
   C. Both A and B
   D. Neither A nor B

   Answer A is incorrect. A faulty clutch pack piston seal will cause the transmission to slip when that clutch pack is engaged.

   Answer B is incorrect. A faulty torque converter stator roller clutch will cause the vehicle to have low power at low speeds due to the lost torque multiplication in the stator.

   Answer C is incorrect. Neither Technician is correct.

   Answer D is correct. Neither Technician is correct. A faulty constant velocity (CV) joint is usually the cause of a vehicle making a clicking noise on turns. A faulty wheel bearing is often the cause of a metallic roar while making turns.

D.   Neither A nor B

Answer A is incorrect. A failed output shaft seal would cause a leak near the output shaft.

Answer B is correct. Only Technician B is correct. A leaking torque converter could cause a transmission fluid leak in the bell housing area.

Answer C is incorrect. Only Technician B is correct.

Answer D is incorrect. Technician B is correct.

6.   Technician A says that only the manufacturer-recommended transmission fluid should be added to an automatic transmission. Technician B says that some manufacturers recommend checking the transmission fluid in park. Who is correct?

TASK A.1.3

   A.   A only

   B.   B only

   C.   Both A and B

   D.   Neither A nor B

Answer A is incorrect. Technician B is also correct.

Answer B is incorrect. Technician A is also correct.

Answer C is correct. Both Technicians are correct. It is vital to use the correct transmission fluid when adding to a transmission. Some manufacturers recommend checking the fluid while idling the engine and with the gearshift in the park position.

Answer D is incorrect. Both Technicians are correct.

7.   Which gear should a technician put the transmission in to create minimum line pressure?

TASK A.1.4

   A.   Manual second

   B.   Manual low

   C.   Park

   D.   Reverse

Answer A is incorrect. The line pressure is sometimes slightly elevated in manual second to increase the holding capacity of the clutches and bands.

Answer B is incorrect. The line pressure in manual low is often elevated in order to increase the holding capacity of the clutches and bands when operating in heavy power conditions.

Answer C is correct. The line pressure in park is usually at a minimum capacity.

Answer D is incorrect. The line pressure in reverse is often elevated to very high levels in order to increase the holding capacity of the drive train while moving in reverse.

8. Technician A says that a pressure test can easily be performed using a scan tool. Technician B says that a pressure test should be performed prior to an initial road test. Who is correct?

A. A only

B. B only

C. Both A and B

D. Neither A nor B

Answer A is incorrect. A pressure gauge set is needed to perform a pressure test on automatic transmissions.

Answer B is incorrect. A pressure test is not usually performed before the initial road test.

Answer C is incorrect. Neither Technician is correct.

Answer D is correct. Neither Technician is correct. A scan tool does not have the capacity to perform a transmission pressure test. The vehicle is usually driven to verify the customer complaint prior to performing a pressure test.

9. A vehicle has a stall speed that is 250 rpm below the standard specification. Technician A says that a slipping torque converter stator could be the cause. Technician B says that a plugged fuel filter could be the cause. Who is correct?

A. A only

B. B only

C. Both A and B

D. Neither A nor B

Answer A is incorrect. Technician B is also correct.

Answer B is incorrect. Technician A is also correct.

Answer C is correct. Both Technicians are correct. A slipping stator could reduce the torque multiplication ability of the torque converter, which would reduce the stall speed. A plugged fuel filter would reduce the engine power, which would reduce the stall speed.

Answer D is incorrect. Both Technicians are correct.

10. Technician A says that a stall test should not be performed until the engine warms up. Technician B says that a stall test should be performed for no longer than five seconds. Who is correct?

A. A only

B. B only

C. Both A and B

D. Neither A nor B

Answer A is incorrect. Technician B is also correct.

Answer B is incorrect. Technician A is also correct.

Answer C is correct. Both Technicians are correct. Stall speed tests should only be performed after the vehicle is warmed up. Stall speed tests should not be performed for longer than five seconds to prevent overheating the transmission fluid.

Answer D is incorrect. Both Technicians are correct.

D.   Neither A nor B

Answer A is incorrect. Technician B is also correct.

Answer B is incorrect. Technician A is also correct.

Answer C is correct. Both Technicians are correct. Many torque converter lockup clutches function by reversing the fluid flow in the torque converter. Most manufacturers do not engage the torque converter clutch below about 35 mph.

Answer D is incorrect. Both Technicians are correct.

12.   A vehicle with an electronically shifted transmission is being road tested. Technician A says that a line pressure gauge can be useful to use during the road test. Technician B says that a shift monitor can be useful during the road test. Who is correct?

TASK A.2.1

A.   A only

B.   B only

C.   Both A and B

D.   Neither A nor B

Answer A is incorrect. It is not a common practice to perform a line pressure test during a road test.

Answer B is correct. Only Technician B is correct. A shift monitor can be a very good diagnostic tool to have during a road test. The shift monitor provides feedback to the technician about when each shift is commanded by the transmission computer.

Answer C is incorrect. Only Technician B is correct.

Answer D is incorrect. Technician B is correct.

13.   Which of the following tools would most likely be used during the road test of a vehicle with an electronic transmission?

TASK A.2.1

A.   Scan tool

B.   Vacuum gauge

C.   Pressure gauge

D.   Micrometer

Answer A is correct. A scan tool is often used during the road test because of the portability of the tool. The scan tool will retrieve data and trouble codes from the transmission computer.

Answer B is incorrect. A vacuum gauge is not typically used during a road test. This tool is sometimes used in the shop when testing vacuum modulators.

Answer C is incorrect. A pressure gauge is not typically used during a road test. This tool is used in the repair shop to test line pressure and governor pressure.

Answer D is incorrect. A micrometer is used to make precision measurements. It would never be used on a road test.

TASK A.2.2

14. The pressure control solenoid connector was left disconnected following a transmission installation. This may cause all of the following EXCEPT:

   A. Transmission will jerk when shifting from park to reverse.

   B. Transmission will jerk when shifting from park to drive.

   C. Transmission will slip at every up-shift.

   D. Transmission will have extremely firm up-shifts.

   Answer A is incorrect. A disconnected pressure control solenoid would cause elevated line pressure, which would cause the transmission to jerk when shifting from park to reverse.

   Answer B is incorrect. A disconnected pressure control solenoid would cause elevated line pressure, which would cause the transmission to jerk when shifting from park to drive.

   Answer C is correct. A disconnected pressure control solenoid would cause elevated line pressure and would not cause the transmission to slip.

   Answer D is incorrect. A disconnected pressure control solenoid would cause elevated line pressure, which would cause the transmission to have extremely firm up-shifts.

TASK A.2.2

15. Which of the following actions would cause the line pressure to increase in a vehicle with an electronic transmission?

   A. Clearing the trouble codes with a scan tool

   B. Disconnecting the pressure control solenoid

   C. Disconnecting the throttle position sensor

   D. Increasing the amperage signal to the pressure control solenoid with a scan tool

   Answer A is incorrect. Clearing the trouble codes with a scan tool would not cause the line pressure to increase.

   Answer B is correct. A disconnected pressure control solenoid would cause the line pressure to rise to the maximum level.

   Answer C is incorrect. Disconnecting the throttle position sensor would cause the check engine light to come on, but would not cause an increase in line pressure.

   Answer D is incorrect. Increasing the amperage signal to the pressure control solenoid would cause the line pressure to drop.

TASK A.2.3

16. A vehicle is being diagnosed for a problem of the engine dying when coming to a stop. After restarting the engine, it again dies when the transmission is shifted into any gear. Technician A says that the torque converter could be the cause. Technician B says that the torque converter clutch solenoid could be the cause. Who is correct?

   A. A only

   B. B only

   C. Both A and B

   D. Neither A nor B

   Answer A is incorrect. Technician B is also correct.

   Answer B is incorrect. Technician A is also correct.

   Answer C is correct. Both Technicians are correct. A stuck torque converter lockup clutch could cause the dying problem. A faulty torque converter clutch solenoid could cause the dying problem by failing to release the lockup circuit.

   Answer D is incorrect. Both Technicians are correct.

and the cruise control because these systems need a vehicle speed input to function.

Answer B is incorrect. A faulty brake switch would affect both the torque converter clutch and the cruise control because these systems need a brake pedal position input to function.

Answer C is correct. A faulty overdrive switch would not likely cause the torque converter clutch or the cruise control to be negatively affected.

Answer D is incorrect. A faulty brake light fuse would cause both the torque converter and the cruise control to be inoperative because the brake switch would not have its supply voltage.

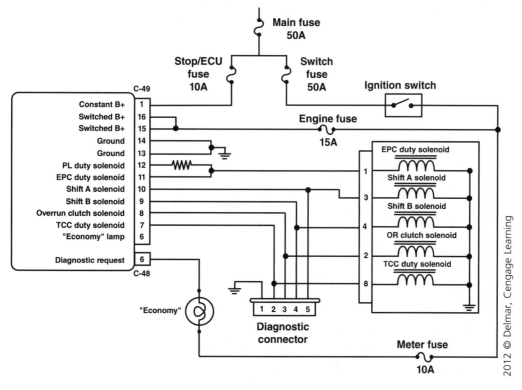

18. Referring to the figure above, Technician A says that all of the solenoids receive ground from the transmission computer. Technician B says that the transmission computer receives power (B+) from the 50 amp main fuse. Who is correct?

TASK A.2.4

A. A only
B. B only
C. Both A and B
D. Neither A nor B

Answer A is incorrect. The solenoids in the schematic are case grounded.

Answer B is correct. Only Technician B is correct. The 50 amp main fuse provides power to the transmission computer at two locations. One location is at connector 49 at pin number 1. The other location is at connector 48 at pin 6.

Answer C is incorrect. Only Technician B is correct.

Answer D is incorrect. Technician B is correct.

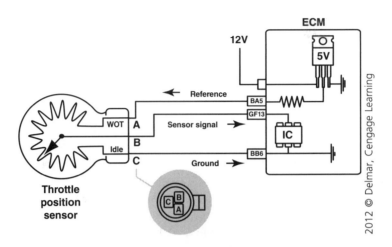

TASK A.2.4

19. Referring to the figure above, Technician A says that the ECM provides a 12 volt supply to the throttle position sensor (TPS). Technician B says that the ECM provides a ground for the TPS at pin BB6 of the ECM. Who is correct?

A. A only

B. B only

C. Both A and B

D. Neither A nor B

Answer A is incorrect. The ECM provides a 5 volt reference voltage to the throttle position sensor.

Answer B is correct. Only Technician B is correct. Pin BB6 is the ECM ground for the TPS. Pin GF13 is the signal wire to the ECM.

Answer C is incorrect. Only Technician B is correct.

Answer D is incorrect. Technician B is correct.

TASK A.2.5

20. The wiring and connections of the charging system should be checked in all of the following ways during generator replacement EXCEPT:

A. Inspect the stator resistance.

B. Inspect the connections for tightness.

C. Inspect the wire insulation for cuts and cracks.

D. Inspect the routing of the wires and harnesses.

Answer A is correct. The stator resistance would not be checked during generator replacement.

Answer B is incorrect. The charging system connections should be checked for tightness to assure that they are secure.

Answer C is incorrect. The charging system wires should be checked for cuts and cracks to assure that they are not damaged.

Answer D is incorrect. The charging system wires should be checked for the correct routing to make sure that they do not get stretched or damaged.

D.   Neither A nor B

Answer A is incorrect. The battery does not need to be recharged. If a battery has at least 9.6 volts at the end of this test, then it passes and can be put back in service.

Answer B is correct. Only Technician B is correct. It is not uncommon for the test lead connectors to get warm during this test.

Answer C is incorrect. Only Technician B is correct.

Answer D is incorrect. Technician B is correct.

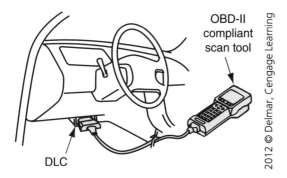

OBD-II
compliant
scan tool

DLC

2012 © Delmar, Cengage Learning

22.   The tool shown in the picture above can be used for all of the following purposes EXCEPT:

A.   Accessing the speed sensor coil resistance
B.   Accessing the transmission computer data list
C.   Accessing the diagnostic trouble codes (DTCs) in the engine computer
D.   Accessing the diagnostic trouble codes (DTCs) in the transmission computer

TASK A.2.6

Answer A is correct. The scan tool shown in the picture would not be used for testing the speed sensor coil resistance. A digital ohmmeter would be used for testing the speed sensor coil.

Answer B is incorrect. The scan tool in the picture could be used for accessing the transmission computer data list.

Answer C is incorrect. The scan tool in the picture could be used for accessing the DTCs in the engine computer.

Answer D is incorrect. The scan tool in the picture could be used for accessing the DTCs in the transmission computer.

23. A vehicle is being diagnosed for a shuddering condition that is most prevalent when the torque converter clutch engages. Which of the following components would be the most likely cause?

    A. Pressure control solenoid

    B. Torque converter

    C. Modulator

    D. Speed sensor

Answer A is incorrect. A pressure control solenoid fault would not likely cause the torque converter clutch to shudder.

Answer B is correct. A faulty torque converter can cause a shuddering condition while driving. A shudder feels like the vehicle is driving over the rough strips at the edge of the road.

Answer C is incorrect. A modulator fault would not likely cause the torque converter clutch to shudder.

Answer D is incorrect. A speed sensor fault would not cause the torque converter clutch to shudder. It could cause the torque converter clutch to be inoperative due to the lack of speed signal.

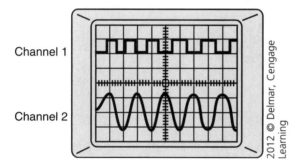

24. Referring to the figure above, Technician A says the pattern on Channel 2 of the oscilloscope is a good signal from a permanent magnet generator (speed sensor). Technician B says the pattern from Channel 1 is a good signal from a throttle position sensor. Who is correct?

    A. A only

    B. B only

    C. Both A and B

    D. Neither A nor B

Answer A is correct. Only Technician A is correct. The analog signal shown on Channel 2 of the oscilloscope would be a good signal from a permanent magnet generator (speed sensor).

Answer B is incorrect. The digital square wave signal shown on Channel 1 of the oscilloscope would not be from a throttle position sensor. This signal would likely originate from a "hall effect" sensor.

Answer C is incorrect. Only Technician A is correct.

Answer D is incorrect. Technician A is correct.

D.   Neither A nor B

Answer A is incorrect. Technician B is also correct.

Answer B is incorrect. Technician A is also correct.

Answer C is correct. Both Technicians are correct. A scan tool can be used to retrieve trouble codes and sensor data from the ECM as well as many other vehicle computers.

Answer D is incorrect. Both Technicians are correct.

26. The transmission shift indicator is not aligned when the vehicle is in park. Technician A says that the manual valve linkage may need to be adjusted. Technician B says that the shift indicator could be out of adjustment. Who is correct?

TASK B.1

A.   A only

B.   B only

C.   Both A and B

D.   Neither A nor B

Answer A is incorrect. Technician B is also correct.

Answer B is incorrect. Technician A is also correct.

Answer C is correct. Both Technicians are correct. A misaligned shift indicator could be caused by either a misadjusted shift linkage or a shift indicator that is out of adjustment.

Answer D is incorrect. Both Technicians are correct.

27. A transaxle is being diagnosed for a problem of early up-shifts. Which of the following faults would most likely cause this problem?

TASK B.2

A.   Broken throttle valve cable

B.   Throttle valve linkage adjusted too tight

C.   Loose band adjustment

D.   Tight band adjustment

Answer A is correct. A broken throttle valve cable would cause the transmission to shift at the earliest point possible due to the transmission not receiving any vehicle load signal.

Answer B is incorrect. A throttle valve cable that is adjusted too tight will cause late up-shifts.

Answer C is incorrect. A loose band adjustment would cause the transmission to slip as the band is engaged.

Answer D is incorrect. A tight band adjustment would cause the transmission to possibly launch in the wrong gear or to even bind up in certain situations.

TASK B.3

28. A transmission is leaking near the rear section of the transmission. Which of the components below is the most likely cause of this leak?

A. Oil pump seal

B. Governor cover seal

C. Manual linkage seal

D. Input speed sensor seal

Answer A is incorrect. A leaking oil pump seal would cause transmission fluid to be present near the bell housing area.

Answer B is correct. The governor cover is near the rear of the transmission, which would cause fluid to be present in this area if the seal was leaking.

Answer C is incorrect. A faulty manual linkage seal would cause the transmission to have fluid on the side of the transmission case near the shift linkage.

Answer D is incorrect. A faulty input speed sensor seal would cause fluid to be leaking near the front section of the transmission case.

TASK B.4

29. All of the following conditions could cause repeat output shaft seal failures EXCEPT:

A. Worn output shaft bushing

B. Overfilled transmission

C. Damaged universal joint at the transmission end of the drive shaft

D. Bent drive shaft

Answer A is incorrect. A worn output shaft bushing could cause the output shaft to have too much movement, which could cause repeat output shaft seal failures.

Answer B is correct. An overfilled transmission would cause the fluid to become aerated, which would cause the transmission to slip.

Answer C is incorrect. A damaged universal joint could cause the drive shaft to vibrate excessively, which could cause repeat failures of the output shaft seal.

Answer D is incorrect. A bent drive shaft could cause repeat output shaft seal failures due to increased movement in the seal area.

TASK B.5

30. Technician A says that the transmission cooler is located in one of the radiator side tanks. Technician B says that leaking transmission cooler lines can be repaired with heater hose and clamps. Who is correct?

A. A only

B. B only

C. Both A and B

D. Neither A nor B

Answer A is correct. Only Technician A is correct. The primary transmission cooler is located in one of the tanks of the engine radiator. Vehicles with a tow package will have an external transmission cooler that is located in front of the radiator and condenser.

Answer B is incorrect. Regular heater hose should not be used to repair transmission cooler hoses. Only high-pressure hose and heavy-duty clamps should be used to make these repairs.

Answer C is incorrect. Only Technician A is correct.

Answer D is incorrect. Technician A is correct.

0.001 C

2012 © Delmar, C

31. What is the measurement reading on the (1 to 2 inch) micrometer above?

   A. 1.270 inches
   B. 1.245 inches
   C. 1.220 inches
   D. 1.230 inches

   TASK B.6,
   C.2.4, C.2.7

   Answer A is incorrect. The reading on the 1 to 2 inch micrometer is 1.245 inches.

   Answer B is correct. The reading on the 1 to 2 inch micrometer is 1.245 inches.

   Answer C is incorrect. The reading on the 1 to 2 inch micrometer is 1.245 inches.

   Answer D is incorrect. The reading on the 1 to 2 inch micrometer is 1.245 inches.

32. A transmission has a broken 1-2 accumulator spring. Technician A says that the transmission will likely shift harshly during the shift from first gear to second gear. Technician B says that the transmission may slip during the shift from second gear to third gear. Who is correct?

   A. A only
   B. B only
   C. Both A and B
   D. Neither A nor B

   TASK B.8

   Answer A is correct. Only Technician A is correct. A broken spring on the 1-2 accumulator would cause a harsh shift from first to second gear.

   Answer B is incorrect. The broken spring on the 1-2 accumulator would not cause a slipping problem. The shift from first to second gear would be harsh and firm.

   Answer C is incorrect. Only Technician A is correct.

   Answer D is incorrect. Technician A is correct.

TASK B.9

33. Technician A says that voltage drop testing a connector needs to be done when the circuit is energized. Technician B says that voltage drop testing a relay needs to be done when the circuit is de-energized. Who is correct?

A. A only

B. B only

C. Both A and B

D. Neither A nor B

Answer A is correct. Only Technician A is correct. Performing a voltage drop test is a good way to determine if a circuit is working correctly. However, this test is only useful when performed on a live circuit.

Answer B is incorrect. All voltage drop tests on electrical components need to be performed when the circuit is energized.

Answer C is incorrect. Only Technician A is correct.

Answer D is incorrect. Technician A is correct.

TASK B.9

34. Technician A states that the scan tool receives data from a connector located on the transmission control module (TCM). Technician B states that the scan tool connects to the data bus using a data link connector (DLC). Who is correct?

A. A only

B. B only

C. Both A and B

D. Neither A nor B

Answer A is incorrect. There is no connector at the TCM to connect the scan tool to.

Answer B is correct. Only Technician B is correct. The scan tool connects to the data bus using a data link connector that is located near the driver's-side area. Once connected, the scan tool is a bi-directional device that can read data, retrieve trouble codes, and send actuator commands to several systems.

Answer C is incorrect. Only Technician B is correct.

Answer D is incorrect. Technician B is correct.

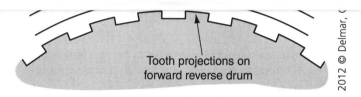

Tooth projections on forward reverse drum

2012 © Delmar, C

35. Referring to the figure above, which electrical test tool could be used to test the voltage pattern that is shown in the picture?

TASK B.9

    A. Voltmeter

    B. Oscilloscope

    C. Ohmmeter

    D. Continuity tester

Answer A is incorrect. A voltmeter would display numbers as the measurement was taken. The picture shows a sine wave signal that could be viewed with an oscilloscope.

Answer B is correct. The pattern in the picture is a sine wave (analog) signal. An oscilloscope would be needed to view the pattern.

Answer C is incorrect. An ohmmeter would be used to measure the resistance of the speed sensor coil, but it would not display the voltage pattern.

Answer D is incorrect. A continuity tester would be used to measure the continuity of the wiring harness.

36. Which of the following methods of wire repair would be most likely used to resist water intrusion in the repair?

TASK B.9

    A. Twist wires together and wrap with electrical tape.

    B. Connect wires with Scotch Lock connectors.

    C. Connect wires by solder and heat shrink.

    D. Connect wires with butt connectors and wrap with electrical tape.

Answer A is incorrect. Electrical tape and wire twisting is not an acceptable wire repair because water can intrude into the connection.

Answer B is incorrect. Scotch Lock connectors have no barrier to repel water from getting into the electrical connection.

Answer C is correct. Wire repairs made by soldering and applying heat shrink will not allow water to enter into the connection.

Answer D is incorrect. Butt connectors and electrical tape will not repel water from entering the electrical connection.

TASK B.10

37. A vehicle is being diagnosed for a bumping noise that occurs when the transmission is shifted to reverse and accelerated. Which of the following components would be the most likely cause?

    A. Chipped planetary gear

    B. Broken rear transmission mount

    C. Transmission band adjusted too loosely

    D. Slipping clutch pack

    Answer A is incorrect. A chipped planetary gear may cause a whining noise at various vehicle speeds.

    Answer B is correct. A broken rear transmission mount could cause a bumping noise as the transmission is shifted to reverse and accelerated. This action would cause the transmission to bump into the floor pan of the vehicle.

    Answer C is incorrect. A loosely adjusted transmission band would cause a transmission slip to occur when the band is engaged.

    Answer D is incorrect. A slipping clutch pack will cause the engine RPM to increase when the clutch pack tries to engage.

TASK C.1.1, C.1.2

38. Which of the following components would be LEAST LIKELY to be removed during a transmission removal process?

    A. Drive shaft

    B. Flywheel dust cover

    C. Rear transmission mount

    D. Engine flywheel

    Answer A is incorrect. The drive shaft would have to be removed during the transmission removal process.

    Answer B is incorrect. The flywheel dust cover would need to be removed in order to gain access to the flexplate and the converter bolts.

    Answer C is incorrect. The rear transmission mount would have to be removed to allow the transmission the clearance to be removed.

    Answer D is correct. The engine flywheel does not need to be removed while taking a transmission out of a vehicle. The bolts securing the torque converter to the flywheel/flexplate have to be removed to allow the torque converter to be separated from the flexplate.

TASK C.1.1

39. A vehicle has a dark-colored fluid leak around the bell housing area. Faults at which of the following locations would most likely cause this problem?

    A. Engine rear main seal

    B. Transmission bottom pan gasket

    C. Axle seal

    D. Speed sensor o-ring

    Answer A is correct. A leaking engine rear main seal would cause the bell housing area to be covered with dark fluid. The technician can determine if the fluid is engine oil or transmission fluid by closely inspecting the color of the fluid.

    Answer B is incorrect. A leaking transmission pan gasket will not cause fluid to be present in the bell housing area.

    Answer C is incorrect. A leaking axle seal would cause fluid to be present near the output shaft area of the transmission.

    Answer D is incorrect. A leaking speed sensor o-ring would cause fluid to be present in the speed sensor area.

to inspect for elongated flexplate mounting holes. These bolt holes would be hidden when the transmission is installed in the vehicle.

Answer B is correct. The ring gear on the flexplate could be closely inspected without removing the transmission. The dust shield would only need to be removed to gain access to this area.

Answer C is incorrect. The transmission would need to be removed from the vehicle in order to inspect for loose flexplate bolts.

Answer D is incorrect. The transmission would need to be removed in order to inspect for a cracked flexplate.

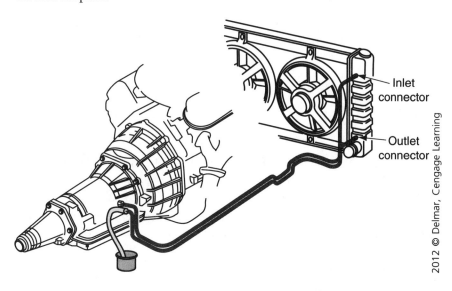

Inlet connector

Outlet connector

2012 © Delmar, Cengage Learning

41. Referring to the figure above, Technician A says the operation in the figure should be performed with the engine at 2,500 rpm. Technician B says that the volume for this test should be approximately one quart of fluid in 20 seconds. Who is correct?

TASK C.1.3

A. A only

B. B only

C. Both A and B

D. Neither A nor B

Answer A is incorrect. The engine should be operating at about 1,000 rpm during the cooler flow test.

Answer B is correct. Only Technician B is correct. The cooler flow test should produce about one quart of transmission fluid in 20 seconds.

Answer C is incorrect. Only Technician B is correct.

Answer D is incorrect. Technician B is correct.

TASK C.3.1

42. A clutch pack needs to be disassembled. Technician A says that the clutch pack snap ring should be removed prior to removing the clutch piston. Technician B says that the clutch drum will need to be thoroughly cleaned when all of the internal parts have been removed. Who is correct?

A. A only

B. B only

C. Both A and B

D. Neither A nor B

Answer A is incorrect. Technician B is also correct.

Answer B is incorrect. Technician A is also correct.

Answer C is correct. Both Technicians are correct. The clutch pack snap ring should be removed prior to removing the clutch piston. The clutch piston is typically the last component to be removed from the drum. The clutch drum will need to be thoroughly cleaned when all of the internal parts have been removed.

Answer D is incorrect. Both Technicians are correct.

TASK C.2.2

43. Technician A says that a new transmission oil pump should be installed dry to prevent contamination. Technician B says that the torque converter hub drives the oil pump on many transmissions. Who is correct?

A. A only

B. B only

C. Both A and B

D. Neither A nor B

Answer A is incorrect. New transmission oil pumps should be lubricated with clean transmission fluid when being installed in order to prime and lubricate the internal components.

Answer B is correct. Only Technician B is correct. The torque converter hub directly drives the oil pump on many transmissions. This allows fluid pressure to be present whenever the engine is running.

Answer C is incorrect. Only Technician B is correct.

Answer D is incorrect. Technician B is correct.

TASK C.2.4

44. Which of the following methods is LEAST LIKELY to be used to adjust end-play on a transmission shaft?

A. Metal thrust washer

B. Brass thrust washer

C. Selective piston plate

D. Metal spacer shim

Answer A is incorrect. Metal thrust washers are sometimes used to adjust the end-play on transmission shafts.

Answer B is incorrect. Brass thrust washers are sometimes used to adjust the end-play on transmission shafts.

Answer C is correct. A selective piston plate is not a common method of adjusting end-play on a transmission shaft.

Answer D is incorrect. Metal spacer shims are sometimes used to adjust the end-play on transmission shafts.

D.   Neither A nor B

Answer A is incorrect. Technician B is also correct.

Answer B is incorrect. Technician A is also correct.

Answer C is correct. Both Technicians are correct. Overheated components should be replaced, as the metal could be weakened by overheating. Anytime that overheating has occurred, the transmission cooler should be inspected and flushed to prevent any future overheating issues.

Answer D is incorrect. Both Technicians are correct.

46.   Technician A says that gasket sealers should be used conservatively to prevent excess sealant from breaking off inside the transmission. Technician B says that some manufacturers do not recommend the use of gasket sealers at all. Who is correct?

TASK C.2.12

A.   A only
B.   B only
C.   Both A and B
D.   Neither A nor B

Answer A is incorrect. Technician B is also correct.

Answer B is incorrect. Technician A is also correct.

Answer C is correct. Both Technicians are correct. If gasket sealers are used in a transmission or transaxle, they should be used sparingly to prevent this substance from getting squeezed out and entering the transmission. Some manufacturers do not allow these types of substance to be used on their transmissions.

Answer D is incorrect. Both Technicians are correct.

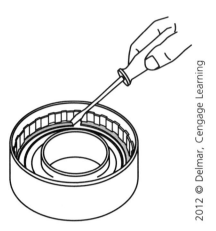

2012 © Delmar, Cengage Learning

TASK C.3.1

47. Which of the following procedures is most likely being performed in the figure above?

  A. Clutch pack apply plate selection

  B. Clutch pack clearance test

  C. Clutch pack snap ring removal

  D. Clutch pack piston return spring service

Answer A is incorrect. A feeler gauge or a dial indicator would be needed to show the clutch pack piston apply plate selection.

Answer B is incorrect. A feeler gauge or a dial indicator would be needed to show the clutch pack clearance test.

Answer C is correct. The picture shows a technician using a screwdriver to remove a clutch pack snap ring. Care should be taken by the technician when performing this activity in order to prevent a cut from the sharp edges of the components.

Answer D is incorrect. A spring compressor would be needed to show the clutch pack piston return spring service.

TASK C.3.2

48. Technician A says the clutch pack clearance may be adjusted by using a selective apply plate. Technician B says that clutch pack clearance may be adjusted by using a selective snap ring. Who is correct?

  A. A only

  B. B only

  C. Both A and B

  D. Neither A nor B

Answer A is incorrect. Technician B is also correct.

Answer B is incorrect. Technician A is also correct.

Answer C is correct. Both Technicians are correct. Selective apply plates and selective snap rings are sometimes used to adjust clutch pack clearance.

Answer D is incorrect. Both Technicians are correct.

49. Referring to the figure above, Technician A says that this test should be performed at each fluid service. Technician B says that this test should be performed at a limited pressure to prevent damage or injury. Who is correct?

    A. A only
    B. B only
    C. Both A and B
    D. Neither A nor B

    TASK C.3.3

    Answer A is incorrect. Air testing a clutch pack should take place at each rebuild of a transmission. This test lets the technician know if the clutch piston will hold pressure.

    Answer B is correct. Only Technician B is correct. Pressure testing clutch packs should be performed at limited air pressure levels to prevent damage or injury during the test.

    Answer C is incorrect. Only Technician B is correct.

    Answer D is incorrect. Technician B is correct.

50. Technician A says that a one-way clutch is energized with hydraulic fluid. Technician B says that a one-way clutch is a holding device. Who is correct?

    A. A only
    B. B only
    C. Both A and B
    D. Neither A nor B

    TASK C.3.4

    Answer A is incorrect. One-way clutches do not need hydraulic fluid applied to them in order to hold members of the planetary gear set. The design of these devices makes them work well for acting as mechanical holding devices.

    Answer B is correct. Only Technician B is correct. One-way clutches used in automatic transmissions operate by holding a member of the planetary gear set.

    Answer C is incorrect. Only Technician B is correct.

    Answer D is incorrect. Technician B is correct.

PREPARATION EXAM 3—ANSWER KEY

1.	C	21.	A	41.	C
2.	C	22.	B	42.	A
3.	C	23.	A	43.	C
4.	B	24.	B	44.	A
5.	C	25.	A	45.	A
6.	C	26.	C	46.	C
7.	B	27.	C	47.	D
8.	C	28.	A	48.	A
9.	D	29.	C	49.	C
10.	B	30.	C	50.	B
11.	A	31.	C		
12.	D	32.	A		
13.	D	33.	C		
14.	A	34.	B		
15.	D	35.	B		
16.	D	36.	B		
17.	A	37.	B		
18.	A	38.	C		
19.	B	39.	C		
20.	D	40.	C		

PREPARATION EXAM 3—EXPLANATIONS

TASK A.1.1

1. A late-model vehicle needs to be road tested to verify a shifting concern. Technician A says that the road test should be performed under the same conditions as described by the customer. Technician B says that the transmission fluid should be checked prior to beginning the road test. Who is correct?

   A. A only
   B. B only
   C. Both A and B
   D. Neither A nor B

   Answer A is incorrect. Technician B is also correct.

   Answer B is incorrect. Technician A is also correct.

   Answer C is correct. Both Technicians are correct. After checking the transmission fluid to verify the condition and level of the fluid, the technician should attempt to drive the vehicle under the same variables as the customer describes on the repair order.

   Answer D is incorrect. Both Technicians are correct.

Answer A is incorrect. A torque converter lockup clutch can cause a shudder in the power train when it is malfunctioning.

Answer B is incorrect. A faulty mass airflow sensor can cause a shudder in the power train. This sensor is a very important input to the engine computer and can cause very noticeable driving concerns if it malfunctions.

Answer C is correct. A chip in the planetary gear would not cause a shudder to occur in the drive train. This problem would only cause a noise in the transmission.

Answer D is incorrect. A faulty engine spark plug can cause a serious misfire in the engine that can feel like a shudder in the drive line.

3. Technician A says the excessive line pressure can cause harsh shifts. Technician B says that low line pressure can cause a shift that causes a flare in the engine RPM. Who is correct?

TASK A.1.2

A. A only

B. B only

C. Both A and B

D. Neither A nor B

Answer A is incorrect. Technician B is also correct.

Answer B is incorrect. Technician A is also correct.

Answer C is correct. Both Technicians are correct. Excessive line pressure will likely cause all of the shifts to be harsh. Low line pressure can cause the transmission to slip when shifting, which will cause the engine RPM to flare up.

Answer D is incorrect. Both Technicians are correct.

4. Technician A says that the engine should be shut off when checking the transmission fluid. Technician B says that some transmissions do not have a dipstick to check the transmission fluid. Who is correct?

TASK A.1.3

A. A only

B. B only

C. Both A and B

D. Neither A nor B

Answer A is incorrect. The engine should be running in order to accurately check the transmission fluid.

Answer B is correct. Only Technician B is correct. It is true that some automatic transmissions do not have a dipstick to use to check the transmission fluid. These transmissions will typically have a vent plug or cap that can be removed to inspect the level of the fluid.

Answer C is incorrect. Only Technician B is correct.

Answer D is incorrect. Technician B is correct.

TASK A.1.3

5. A vehicle is being diagnosed that has extremely burned and discolored transmission fluid. Which of the following problems would be LEAST LIKELY to cause this condition with the fluid?

   A. Transmission cooler restricted

   B. Misadjusted band

   C. Faulty transmission computer

   D. Leaking clutch pack piston

   Answer A is incorrect. A restricted transmission cooler would cause the transmission fluid to overheat, which would cause it to become burned and discolored.

   Answer B is incorrect. A band that is adjusted too loose will cause the band to slip when applied, which would cause the fluid to become burned and discolored.

   Answer C is correct. A faulty transmission computer would not likely cause the transmission fluid to become burned and discolored.

   Answer D is incorrect. A leaking clutch pack piston would cause a clutch pack to slip when applied. This slipping of the clutch pack would cause the fluid to become burned and discolored.

TASK A.1.4

6. Technician A says that the governor pressure should be approximately 50 psi at 50 mph. Technician B says that governor pressure should be near zero when the vehicle is stopped. Who is correct?

   A. A only

   B. B only

   C. Both A and B

   D. Neither A nor B

   Answer A is incorrect. Technician B is also correct.

   Answer B is incorrect. Technician A is also correct.

   Answer C is correct. Both Technicians are correct. The governor valve is the device that creates a fluid pressure signal to send to the valve body. The pressure should be close to the vehicle speed, which means that the pressure should be zero at a stop and at about 50 psi at 50 miles per hour.

   Answer D is incorrect. Both Technicians are correct.

TASK A.1.4

7. A pressure test can reveal all of the following results EXCEPT:

   A. Clutch application pressure

   B. Band pressure

   C. Mainline pressure

   D. Governor pressure

   Answer A is incorrect. A pressure test can show the pressure that is sent to a clutch pack.

   Answer B is correct. A pressure test will not test the pressure that the band creates when holding a drum.

   Answer C is incorrect. A pressure test can show the mainline pressure that the transmission pump is creating.

   Answer D is incorrect. A pressure test can show the governor pressure. The pressure should be close to the vehicle speed, which means that the pressure should be zero at a stop and at about 50 psi at 50 miles per hour.

D. Neither A nor B

Answer A is incorrect. Technician B is also correct.

Answer B is incorrect. Technician A is also correct.

Answer C is correct. Both Technicians are correct. Anything that can cause the engine power to be reduced can cause the stall speed to be reduced. Low fuel pressure as well as a plugged exhaust system can cause reduced stall speed.

Answer D is incorrect. Both Technicians are correct.

9. Which of the following transmission components is LEAST LIKELY to be tested during a stall test?

TASK A.1.5

   A. Band

   B. Clutch pack

   C. Torque converter stator

   D. Governor valve

Answer A is incorrect. A band is tested during a stall test. If a band is slipping, the stall speed will be higher than the specification.

Answer B is incorrect. A clutch pack is tested during a stall test. If a clutch pack is slipping, the stall speed will be higher than the specification.

Answer C is incorrect. The torque converter stator is tested during a stall test. If the stator does not lock during the stall test, then the stall speed will be lower than the specification.

Answer D is correct. The governor valve is the device that creates a fluid pressure signal to send to the valve body. A stall test will not test the governor valve. A pressure test will need to be performed to test the governor valve.

10. Technician A says that the engine RPM should drop 25 rpm when the torque converter clutch engages. Technician B says that the torque converter clutch should disengage when the vehicle is slowed down below the engagement speed. Who is correct?

TASK A.1.6

   A. A only

   B. B only

   C. Both A and B

   D. Neither A nor B

Answer A is incorrect. The RPM should drop about 150 to 200 rpm when the torque converter clutch engages.

Answer B is correct. Only Technician B is correct. The torque converter clutch should disengage when the vehicle slows down below the converter clutch engagement speed.

Answer C is incorrect. Only Technician B is correct.

Answer D is incorrect. Technician B is correct.

11. Technician A says that the torque converter lockup clutch is not typically functional on a cold engine. Technician B says that the torque converter lockup clutch creates high levels of heat in the torque converter. Who is correct?

    A. A only
    B. B only
    C. Both A and B
    D. Neither A nor B

    Answer A is correct. Only Technician A is correct. The torque converter lockup clutch is not usually commanded to engage when the engine is cold. This strategy reduces the drive train load on the engine when it is still warming up.

    Answer B is incorrect. The torque converter clutch does not create extra heat in the torque converter. In fact, the converter clutch causes the vortex fluid flow to stop when it engages.

    Answer C is incorrect. Only Technician A is correct.

    Answer D is incorrect. Technician A is correct.

12. Which of the following tools would most likely be used during the road test of a vehicle with an electronic transmission?

    A. Dial indicator
    B. Vacuum gauge
    C. Pressure gauge
    D. Scan tool

    Answer A is incorrect. A dial indicator is used for precision measurement and would not be used on a road test.

    Answer B is incorrect. A vacuum gauge is used to measure engine vacuum and would not likely be used on a road test.

    Answer C is incorrect. A pressure gauge is used to measure hydraulic pressure in the transmission and would not likely be used on a road test.

    Answer D is correct. A scan tool is often used during a road test to monitor data and trouble codes at the transmission computer.

13. A vehicle with an electronic transmission is being road tested. All of the following items should be observed during the road test EXCEPT:

    A. Correct up-shift speeds
    B. Unusual drive train noises
    C. Unusual downshift vibration
    D. Fuel economy

    Answer A is incorrect. The technician should observe the shift points for all up-shift actions of the transmission during the road test.

    Answer B is incorrect. The technician should listen for unusual drive train noises during the road test.

    Answer C is incorrect. The technician should observe the downshift actions of the transmission to see if any unusual vibrations or noises occur.

    Answer D is correct. The technician will not drive the vehicle long enough to measure the fuel economy on a road test.

D.   Neither A nor B

Answer A is correct. Only Technician A is correct. A scan tool can be used to signal the pressure control solenoid to raise the pressure during a diagnosis.

Answer B is incorrect. Putting the transmission in neutral will not cause the transmission pressure to move to a maximum point.

Answer C is incorrect. Only Technician A is correct.

Answer D is incorrect. Technician A is correct.

15.   A vehicle with an electronic transmission has been tested for transmission pump output pressure. The pressure test results are below the specification in all ranges. Which of the following conditions would most likely cause this problem?

TASK A.2.2

A.   Slipping torque converter stator

B.   Front pump seal leaking externally

C.   Loose torque converter mounting bolt

D.   Transmission filter not properly secured to the pump

Answer A is incorrect. A slipping torque converter stator will cause low power when taking off in the vehicle.

Answer B is incorrect. A front pump seal will cause transmission fluid to be present in the bell housing area.

Answer C is incorrect. A loose torque converter mounting bolt may cause a clicking noise.

Answer D is correct. A problem with the transmission filter can cause reduced pressures in all ranges.

16.   A vehicle is being diagnosed for a problem of the engine dying when coming to a stop. After restarting the engine, it again dies when the transmission is shifted into any gear. Which of the following is most likely to be the cause?

TASK A.2.3

A.   An open transmission temperature sensor

B.   A shorted speed sensor

C.   A faulty generator

D.   Stuck torque converter clutch solenoid

Answer A is incorrect. An open transmission temperature sensor will cause a trouble code in the transmission computer.

Answer B is incorrect. A shorted speed sensor will cause a trouble code in the transmission computer. In addition, the transmission may be stuck in "limp in" mode due to the lack of a speed sensor signal.

Answer C is incorrect. A faulty generator will cause the voltage level to be lowered, which can eventually cause the engine to stall and fail to crank because the battery will be discharged.

Answer D is correct. A stuck torque converter clutch solenoid can cause the torque converter clutch to remain engaged, which will cause the engine to die when put in gear.

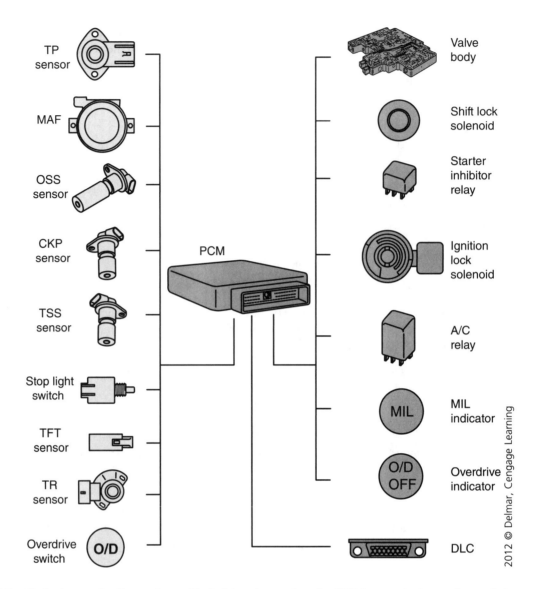

17. Referring to the figure above, Technician A says that the PCM controls the engine and the electronic transmission. Technician B says that the PCM does not control the shift interlock system. Who is correct?

TASK A.2.4

A.   A only

B.   B only

C.   Both A and B

D.   Neither A nor B

Answer A is correct. Only Technician A is correct. The schematic shows that the PCM handles the engine and transmission functions.

Answer B is incorrect. The schematic shows that the PCM does control the shift interlock solenoid.

Answer C is incorrect. Only Technician A is correct.

Answer D is incorrect. Technician A is correct.

Answer A is correct. Only Technician A is correct. A failed brake switch can cause the torque converter clutch to not work. This system uses the brake switch to turn off the torque converter clutch when the brakes are applied.

Answer B is incorrect. A restricted transmission filter would not likely cause the torque converter clutch to be inoperative. The transmission would not pull at all if the filter was stopped up.

Answer C is incorrect. Only Technician A is correct.

Answer D is incorrect. Technician A is correct.

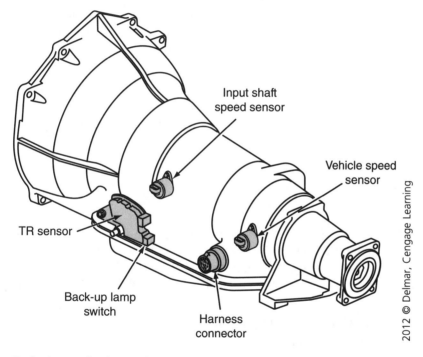

Input shaft speed sensor

Vehicle speed sensor

TR sensor

Back-up lamp switch

Harness connector

2012 © Delmar, Cengage Learning

19. Referring to the figure above, Technician A says that the input shaft speed sensor is a thermistor that varies its resistance as the vehicle speed changes. Technician B says that the input shaft speed sensor is used on electronic transmissions as an input to calculate the gear ratio to determine when a clutch pack may be slipping. Who is correct?

TASK A.2.4

A. A only
B. B only
C. Both A and B
D. Neither A nor B

Answer A is incorrect. The input shaft speed sensor is a permanent magnet (PM) generator that creates a voltage signal as the input shaft rotates.

Answer B is correct. Only Technician B is correct. The input shaft speed sensor is used on electronic transmissions to allow the transmission computer to calculate gear ratios. Having this information allows the transmission computer to sense when the transmission is slipping.

Answer C is incorrect. Only Technician B is correct.

Answer D is incorrect. Technician B is correct.

TASK A.2.5

20. A digital battery tester would be most likely used for which of the following purposes?

    A. Measuring resistance of control modules

    B. Testing starter current draw

    C. Measuring voltage of control modules

    D. Testing the impedance of the battery

    Answer A is incorrect. A technician should never attempt to measure the resistance of a control module. A digital ohmmeter could be used to measure resistance in electronic sensors and circuits after they are disconnected from the circuit.

    Answer B is incorrect. An electronic tool such as a digital multi-meter (DMM) used with an amp clamp, or possibly an electrical system tester that utilizes an amp clamp, can be used to test starter draw.

    Answer C is incorrect. A high-impedance DMM set to read DC volts could be used to read voltage levels of control module circuits.

    Answer D is correct. A digital battery tester can be used to test the impedance of a vehicle battery. It is the recommended tool in the industry because it can accurately test a battery even if it just needs to be charged. Test results from this tester are very conclusive on what needs to be done with the battery.

TASK A.2.5

21. Technician A says that the voltage drop on the positive battery cable should be less than 0.5 volts while cranking the engine. Technician B says that the voltage drop on the negative battery cable should be less than 1.5 volts while cranking the engine. Who is correct?

    A. A only

    B. B only

    C. Both A and B

    D. Neither A nor B

    Answer A is correct. Only Technician A is correct. The voltage drop should be less than 0.5 volts while cranking the engine. If the voltage is higher than this, the cable has a problem.

    Answer B is incorrect. The voltage drop on the negative battery cable should be less than 0.5 volts while cranking the engine. A voltage drop of 1.5 volts would indicate a problem in the cable.

    Answer C is incorrect. Only Technician A is correct.

    Answer D is incorrect. Technician A is correct.

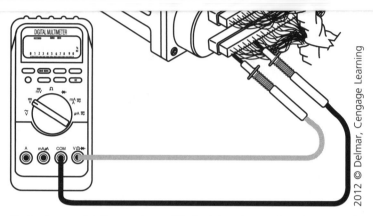

2012 © Delmar, Cengage Learning

22. Referring to the figure above, Technician A says that the meter is measuring the resistance of the module. Technician B says that all meters used in this manner need to be a high-impedance design. Who is correct?

TASK A.2.6

   A. A only

   B. B only

   C. Both A and B

   D. Neither A nor B

   Answer A is incorrect. A technician would never measure the resistance of a module because damage to the module would likely occur.

   Answer B is correct. Only Technician B is correct. A high-impedance meter would be necessary when measuring voltage on any circuit that involves an electronic module.

   Answer C is incorrect. Only Technician B is correct.

   Answer D is incorrect. Technician B is correct.

23. A vehicle is being diagnosed for a shuddering condition that is most prevalent when the torque converter clutch engages. This problem could be caused by all of the following components EXCEPT:

TASK A.2.6

   A. Pressure control solenoid

   B. Pulse-width modulated torque converter clutch solenoid

   C. Torque converter clutch apply plate

   D. Torque converter clutch friction disc

   Answer A is correct. A pressure control solenoid would not cause a shudder in the torque converter clutch. The transmission line pressure would be affected if the solenoid failed.

   Answer B is incorrect. A fault in the modulated torque converter clutch solenoid could cause the torque converter to shudder when it engages.

   Answer C is incorrect. A damaged torque converter clutch apply plate could cause the torque converter clutch to shudder when it engages.

   Answer D is incorrect. A damaged torque converter clutch friction disc could cause the torque converter clutch to shudder when it engages.

TASK A.2.7

24. A problem with the following electronic transmission components could cause the transmission to enter "limp mode" EXCEPT:

    A. Shift solenoid A

    B. Overdrive switch

    C. Shift solenoid B

    D. Input shaft speed sensor

    Answer A is incorrect. A fault in any of the shift solenoids can cause the transmission to go into "limp mode."

    Answer B is correct. The overdrive switch would not cause the transmission to enter "limp mode."

    Answer C is incorrect. A fault in any of the shift solenoids can cause the transmission to go into "limp mode."

    Answer D is incorrect. A fault in the input shaft speed sensor can cause the transmission to go into "limp mode."

TASK A.2.7

25. A vehicle with an electronic transmission will not move in forward or reverse. Which of the following could cause this problem?

    A. Broken input shaft

    B. Faulty input shaft speed sensor

    C. Faulty shift solenoid

    D. Transmission fluid level is one quart overfilled

    Answer A is correct. A broken input shaft would cause the transmission to be totally inoperative.

    Answer B is incorrect. A faulty input speed sensor would cause the transmission to be in "limp mode," but it would still have reverse and one forward gear.

    Answer C is incorrect. A faulty shift solenoid would cause the transmission to be in "limp mode," but it would still have reverse and one forward gear.

    Answer D is incorrect. A transmission that is overfilled with transmission fluid may slip some, but it would still move the vehicle.

TASK B.1

26. Which of the components below is used as an input to the transmission computer for gearshift position?

    A. MAP sensor

    B. Back-up switch

    C. Range sensor

    D. Manual valve

    Answer A is incorrect. The MAP sensor is used as an engine load input to the transmission computer. The data from the MAP sensor will likely be sent to the transmission computer over the data bus network from the engine computer.

    Answer B is incorrect. The back-up switch would not provide data about all of the gearshift positions.

    Answer C is correct. The range sensor provides gearshift data to the transmission computer.

    Answer D is incorrect. The manual valve is the physical device that is connected to the gearshift. However, this device is just a mechanical valve that routes fluid in the valve body.

D. Neither A nor B

Answer A is incorrect. Technician B is also correct.

Answer B is incorrect. Technician A is also correct.

Answer C is correct. Both Technicians are correct. The throttle valve cable is the load input to the valve body. If this cable is not adjusted correctly, the shift points will be affected. If the cable is too loose, the shift points will be early. If the cable is adjusted too tight, the shift points will be too late.

Answer D is incorrect. Both Technicians are correct.

28. A transmission is leaking near the rear section of the transmission. All of the components below could cause this problem EXCEPT:

    A. Oil pump seal
    B. Extension housing gasket
    C. Output shaft seal
    D. Output speed sensor seal

TASK B.3

Answer A is correct. A leaking oil pump seal will cause fluid to be present in the bell housing area of the transmission.

Answer B is incorrect. A leaking extension housing gasket will cause fluid to be present near the rear of the transmission.

Answer C is incorrect. A leaking output shaft seal will cause fluid to be present near the rear of the transmission.

Answer D is incorrect. A leaking output speed sensor seal will cause fluid to be present near the rear of the transmission.

29. A transmission is being diagnosed for a worn extension housing bushing. Technician A says that the old bushing could be compared to the new bushing during the inspection. Technician B says that the drive yoke should be closely checked for deep scratches during the inspection. Who is correct?

    A. A only
    B. B only
    C. Both A and B
    D. Neither A nor B

TASK B.4

Answer A is incorrect. Technician B is also correct.

Answer B is incorrect. Technician A is also correct.

Answer C is correct. Both Technicians are correct. When inspecting the extension housing bushing, the technician should compare the old bushing with a new one. In addition, the technician should closely inspect the drive yoke for scratches, which would require the drive yoke to be replaced.

Answer D is incorrect. Both Technicians are correct.

TASK B.5

30. A vehicle is in the shop for an overheating problem. During the inspection the technician finds that upper radiator hose collapses after the engine cools down but moves back to normal when the radiator cap is removed. What is the most likely cause for this problem?

    A. Faulty water pump
    B. Malfunctioning cooling fan
    C. Faulty radiator cap
    D. Faulty thermostat

    Answer A is incorrect. A faulty water pump could fail by leaking or by not pumping enough coolant, but would not cause the conditions described in this question.

    Answer B is incorrect. A malfunctioning cooling fan could cause overheating at slow speeds or while parked, but would not cause the conditions described in this question.

    Answer C is correct. The vacuum valve in the radiator cap is not allowing the coolant to return to the radiator when the temperature cools back down. This problem is signaled by the upper radiator hose collapsing.

    Answer D is incorrect. A faulty thermostat can stick open and cause overcooling as well as sticking closed and cause overheating, but it would not cause the conditions described in this question.

TASK B.7

31. Technician A says that the valve body is typically tightened up with an inch/pound torque wrench. Technician B says that the valve body should be tightened to a specified torque in a uniform manner. Who is correct?

    A. A only
    B. B only
    C. Both A and B
    D. Neither A nor B

    Answer A is incorrect. Technician B is also correct.

    Answer B is incorrect. Technician A is also correct.

    Answer C is correct. Both Technicians are correct. The valve body is a very critical and fragile part of an automatic transmission. Extreme care should be used when tightening the bolts for the valve body. An inch/pound torque wrench is often used to complete the tightening process. A uniform tightening sequence should be used when tightening the valve body bolts.

    Answer D is incorrect. Both Technicians are correct.

TASK B.9

32. Technician A says that intermittent electrical signals can sometimes be diagnosed by using an oscilloscope. Technician B says that a broken tooth on a speed sensor reluctor can sometimes be diagnosed by using an ohmmeter. Who is correct?

    A. A only
    B. B only
    C. Both A and B
    D. Neither A nor B

    Answer A is correct. Only Technician A is correct. An oscilloscope displays voltage over time on a digital screen. Viewing signals this way allows the technician to pick up electrical glitches and intermittent signals that happen very quickly.

    Answer B is incorrect. Testing the resistance of a sensor will not typically pick up something like a broken tooth on a speed sensor. An oscilloscope is an effective tool for finding this type of problem.

    Answer C is incorrect. Only Technician A is correct.

    Answer D is incorrect. Technician A is correct.

Answer A is incorrect. A digital voltmeter is an acceptable tool to use when testing transmission sensors as long as the meter has high impedance.

Answer B is incorrect. An oscilloscope is an acceptable tool to use when testing transmission sensors. These tools are especially good for testing sensors for intermittent problems.

Answer C is correct. A test light should never be used when testing any circuit that has a connection to a control module. The test light has very low impedance (internal resistance), which could damage sensitive electronic circuits.

Answer D is incorrect. A digital ohmmeter is a common tool that is used when testing transmission sensors.

34. Which of the following definitions best describes a terminal?

TASK B.9

    A. A magnetic switch

    B. A metallic component used to tie electrical circuits together

    C. A plastic housing used to hold terminals as well as to plug into electrical items

    D. A joint where two or more electrical circuits connect to each other

Answer A is incorrect. A magnetic switch is usually defined as a relay.

Answer B is correct. A terminal is the metallic component that is used to join electrical circuits.

Answer C is incorrect. A plastic housing that is used to hold terminals as well as to plug into electrical items is defined as a connector.

Answer D is incorrect. A splice is a joint where two or more electrical circuits connect to each other.

35. Which of the following conditions would be the LEAST LIKELY cause of a thumping sound that happens when accelerating quickly?

TASK B.10

    A. Broken engine mount

    B. Faulty axle bearing

    C. Broken transmission mount

    D. Faulty universal joint

Answer A is incorrect. A broken engine mount can cause a thumping sound while accelerating quickly due to increased movement of the engine.

Answer B is correct. A faulty axle bearing will cause a roaring sound that increases as the vehicle is making turns.

Answer C is incorrect. A broken transmission mount can cause a thumping sound while accelerating quickly due to increased movement of the transmission housing.

Answer D is incorrect. A faulty universal joint can cause a thumping sound while accelerating quickly due to increased movement in the driveline.

36. Which gear is most likely to be selected when checking the transmission fluid level?

    A. Manual low
    B. Park
    C. Reverse
    D. Overdrive

    Answer A is incorrect. The transmission should not be in manual low gear when checking the transmission fluid.

    Answer B is correct. The transmission fluid level is checked in park gear on many late-model vehicles.

    Answer C is incorrect. The transmission should not be in reverse gear when checking the transmission fluid.

    Answer D is incorrect. The transmission should not be in overdrive gear when checking the transmission fluid.

37. Which type of transmission fluid should be used in an automatic transmission?

    A. Dexron 4
    B. Manufacturer recommended type
    C. ATF+4
    D. Dexron with Mercon

    Answer A is incorrect. Dexron 4 is used in some automatic transmissions, but the owner's manual should be referenced to see exactly the type of fluid that should be used.

    Answer B is correct. Only the fluid that is recommended by the manufacturer should be used in automatic transmissions. Using the wrong fluid can cause shifting problems in many late-model transmissions.

    Answer C is incorrect. ATF+4 is used in some automatic transmissions, but the owner's manual should be referenced to see the exact type of fluid that should be used.

    Answer D is incorrect. Dexron with Mercon is used in some automatic transmissions, but the owner's manual should be referenced to see the exact type of fluid that should be used.

38. A transaxle needs to be removed from a late-model car. Technician A installs an engine support fixture before removing the engine subframe. Technician B removes the air cleaner before removing the upper bell housing bolts. Who is correct?

    A. A only
    B. B only
    C. Both A and B
    D. Neither A nor B

    Answer A is incorrect. Technician B is also correct.

    Answer B is incorrect. Technician A is also correct.

    Answer C is correct. Both Technicians are correct. Before removing a transaxle, some type of engine support fixture should be attached to the top side of the engine in order to maintain support when the transaxle is removed. It is also a good idea to remove some of the upper engine components to gain access to the bell housing bolts.

    Answer D is incorrect. Both Technicians are correct.

Answer A is incorrect. Technician B is also correct.

Answer B is incorrect. Technician A is also correct.

Answer C is correct. Both Technicians are correct. When installing a transmission, it is advisable to make sure that the torque converter is fully seated into the transaxle. In addition, the engine dowel pins should be checked to assure that they are seated and in place.

Answer D is incorrect. Both Technicians are correct.

40. Technician A says that the torque converter position should be noted prior to removing from the transmission. Technician B says that the torque converter should be thoroughly installed into the transmission prior to installing the transmission. Who is correct?

TASK C.1.2

A. A only

B. B only

C. Both A and B

D. Neither A nor B

Answer A is incorrect. Technician B is also correct.

Answer B is incorrect. Technician A is also correct.

Answer C is correct. Both Technicians are correct. It is advisable to inspect the position of the old torque converter prior to removing it from the transmission. Doing this will give a point of reference when installing the converter back into the transmission. It is very important to fully seat the torque converter into the transmission case before installing the transmission into the vehicle.

Answer D is incorrect. Both Technicians are correct.

41. Technician A says that an external transmission cooler should be used on any vehicle that is loaded heavily on a regular basis. Technician B says that some transmission coolers have small baffles to increase the heat transfer. Who is correct?

TASK C.1.3

A. A only

B. B only

C. Both A and B

D. Neither A nor B

Answer A is incorrect. Technician B is also correct.

Answer B is incorrect. Technician A is also correct.

Answer C is correct. Both Technicians are correct. External transmission coolers are used to increase the fluid cooling capacity on vehicles that are typically loaded heavily. These coolers are located in front of the radiator and condenser. Some transmission coolers have small baffles in order to assist the heat transfer ability of the cooler.

Answer D is incorrect. Both Technicians are correct.

42. Which of the following methods would be most likely used in cleaning the transmission case after the internal parts have been removed?

    A.   Parts solvent and compressed air

    B.   Air-powered die grinder with a buffing wheel

    C.   Aerosol cleaner and a paper shop towel

    D.   Parts solvent and a shop rag

    Answer A is correct. Parts solvent and compressed air work well when cleaning the transmission case.

    Answer B is incorrect. A die grinder should never be used on a transmission/case because the small particles generated by this tool will get lodged into places that will cause damage.

    Answer C is incorrect. A paper shop towel should never be used when cleaning a transmission case. Debris from the towel can get lodged in locations that will cause future problems.

    Answer D is incorrect. A shop rag should never be used to clean a transmission case because lint from the rag can get lodged in locations that will cause future problems.

43. Which of the following tools would be LEAST LIKELY to be used when measuring a gear-type oil pump assembly?

    A.   Feeler gauge

    B.   Straightedge

    C.   Ruler

    D.   Micrometer

    Answer A is incorrect. A feeler gauge is often used when measuring the clearance between pump gears and crescents.

    Answer B is incorrect. A straightedge is used along with a feeler gauge to check for the flatness of the pump cover and housing.

    Answer C is correct. A ruler would not be a useful tool when measuring an oil pump for wear.

    Answer D is incorrect. A micrometer can be used to accurately measure diameters and thicknesses of the pump components.

44. Technician A says that bearing preload is sometimes measured by testing the turning torque on a shaft that has been assembled. Technician B says that backlash is measured with a micrometer. Who is correct?

    A.   A only

    B.   B only

    C.   Both A and B

    D.   Neither A nor B

    Answer A is correct. Only Technician A is correct. It is common to measure the turning torque on an assembled shaft. This test is a method of checking the bearing preload.

    Answer B is incorrect. A dial indicator is needed to measure the backlash of a gear set.

    Answer C is incorrect. Only Technician A is correct.

    Answer D is incorrect. Technician A is correct.

D. Neither A nor B

Answer A is correct. Only Technician A is correct. Any plastic thrust washers that have metal embedded in them should be replaced, as the metal could damage adjacent mating surfaces.

Answer B is incorrect. Plastic parts that are bad should be replaced with new plastic parts.

Answer C is incorrect. Only Technician A is correct.

Answer D is incorrect. Technician A is correct.

46. Technician A says that transmission shafts should be checked for scoring on all areas that ride next to a bushing. Technician B says that transmission shafts should be checked for scoring on all areas that ride next to a bearing. Who is correct?

TASK C.2.5

A. A only
B. B only
C. Both A and B
D. Neither A nor B

Answer A is incorrect. Technician B is also correct.

Answer B is incorrect. Technician A is also correct.

Answer C is correct. Both Technicians are correct. Transmission shafts that ride on a bushing or a bearing should always be closely inspected for wear and scratches.

Answer D is incorrect. Both Technicians are correct.

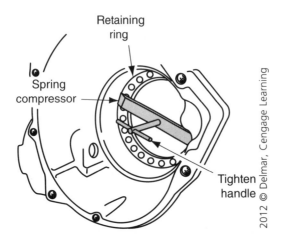

47. Which of the following procedures is most likely being performed in the figure above?

    A. Clutch pack apply plate selection
    B. Clutch pack clearance test
    C. Clutch pack snap ring removal
    D. Clutch pack piston return spring service

    TASK C.3.1

    Answer A is incorrect. A dial indicator or a feeler gauge would be needed to assist in the clutch pack apply plate selection.

    Answer B is incorrect. A dial indicator or a feeler gauge would be needed to assist in the clutch pack clearance test.

    Answer C is incorrect. A screwdriver would be needed to remove the clutch pack snap ring.

    Answer D is correct. The tool shown in the picture is used to compress the piston return spring assembly.

48. All of the following methods are sometimes used to adjust the clutch pack clearance EXCEPT:

    A. Selective piston assembly
    B. Selective apply plate
    C. Adding an extra steel plate
    D. Selective snap ring

    TASK C.3.2

    Answer A is correct. Selective pistons are not used to adjust clutch pack clearance.

    Answer B is incorrect. Selective apply plates are sometimes used to adjust clutch pack clearance.

    Answer C is incorrect. Although this is not a common method, adding an extra steel plate can be a method to adjust clutch pack clearance.

    Answer D is incorrect. Selective snap rings are sometimes used to adjust the clutch pack clearance.

TASK C.3.3

49. Referring to the figure above, Technician A says that this test should be performed prior to assembling the clutch pack into the transmission. Technician B says that this test will expose a problem in the clutch pack piston and seal area. Who is correct?

    A. A only
    B. B only
    C. Both A and B
    D. Neither A nor B

    Answer A is incorrect. Technician B is also correct.

    Answer B is incorrect. Technician A is also correct.

    Answer C is correct. Both Technicians are correct. Air tests on a clutch pack should be performed when possible to see if the clutch pack is assembled correctly. If a seal is damaged in the clutch pack, then air can be heard leaking during the test.

    Answer D is incorrect. Both Technicians are correct.

50. A transmission is being diagnosed for a problem of slipping in "OD" first gear. However, the transmission does not slip when the manual low position is used on the gearshift. What is the most likely cause of this problem?

    TASK C.3.4

    A. The torque converter impeller is weak.
    B. The low gear roller clutch is slipping.
    C. The forward clutch pack assembly has a blown piston seal.
    D. The low gear servo is leaking.

    Answer A is incorrect. A weak torque converter impeller would affect the transmission operation in all ranges.

    Answer B is correct. A slipping low gear roller clutch could cause the problem. The fact that the vehicle pulls when in manual low provides the information that exposes the low roller clutch.

    Answer C is incorrect. A blown piston seal in the forward clutch pack assembly would not be affected by shifting to manual low gear.

    Answer D is incorrect. A leaking low gear servo would loosely apply the low gear band and would not hold very well.

PREPARATION EXAM 4—ANSWER KEY

1.	C	21.	B	41.	B
2.	D	22.	A	42.	C
3.	D	23.	C	43.	A
4.	A	24.	B	44.	B
5.	B	25.	D	45.	B
6.	B	26.	B	46.	D
7.	B	27.	C	47.	A
8.	A	28.	B	48.	B
9.	D	29.	A	49.	B
10.	A	30.	C	50.	C
11.	A	31.	B		
12.	A	32.	B		
13.	B	33.	D		
14.	C	34.	C		
15.	A	35.	B		
16.	B	36.	A		
17.	D	37.	C		
18.	C	38.	B		
19.	C	39.	B		
20.	C	40.	C		

PREPARATION EXAM 4—EXPLANATIONS

TASK A.1.1

1. Technician A says that a broken transmission mount assembly could cause a knocking noise when the vehicle is accelerated quickly during a road test. Technician B says that a missing retaining bolt for a transmission mount could cause a knocking noise when the vehicle is shifted into reverse during a road test. Who is correct?

    A. A only
    B. B only
    C. Both A and B
    D. Neither A nor B

    Answer A is incorrect. Technician B is also correct.

    Answer B is incorrect. Technician A is also correct.

    Answer C is correct. Both Technicians are correct. A broken transmission mount or a missing bolt for a transmission mount could cause a knocking noise during a road test. Either of these problems would allow the transmission to move more than it is supposed to when torque is applied to the vehicle.

    Answer D is incorrect. Both Technicians are correct.

D.   Neither A nor B

Answer A is incorrect. A torque converter clutch that remains engaged too long would cause the engine to die when the vehicle was slowing to a stop.

Answer B is incorrect. A speed sensor that is not functioning all of the time could cause the vehicle to not shift correctly or to go into "limp mode."

Answer C is incorrect. Neither Technician is correct.

Answer D is correct. Neither Technician is correct. The most likely cause of a vehicle that slips when making a sharp turn would be low transmission fluid level.

3. Which of the following drive train components would be most likely to cause a roaring noise that increases when the wheels are turned to the right?

TASK A.1.2

A.   Right-front wheel bearing
B.   Loose torque converter bolt
C.   Transmission oil pump
D.   Left-front wheel bearing

Answer A is incorrect. The right-front wheel bearing would cause the roaring noise to increase when turning the wheels to the left.

Answer B is incorrect. A loose torque converter bolt could cause a knocking noise in the bell housing area.

Answer C is incorrect. A transmission oil pump could cause a continuous whining sound that would not be changed by changing the wheel direction.

Answer D is correct. A left-front wheel bearing would cause a roaring sound that increases when the wheels are turned to the right. Turning to the right would increase the mechanical load on the left side of the vehicle.

4. All of the following faults could cause a clicking noise when the vehicle is put into gear and quickly accelerated while holding the brakes EXCEPT:

TASK A.1.2

A.   Worn spider gears
B.   Cracked flexplate
C.   Loose torque converter bolts
D.   Loose flywheel bolts

Answer A is correct. Worn spider gears could cause a roaring sound in the driveline when making turns.

Answer B is incorrect. A cracked flexplate could cause a clicking noise when the vehicle is accelerated due to the increased load on the flexplate under these conditions.

Answer C is incorrect. Loose torque converter bolts could cause a clicking noise when the vehicle is accelerated due to the increased load on the flexplate under these conditions.

Answer D is incorrect. Loose flywheel bolts could cause a clicking noise when the vehicle is accelerated due to the increased load on the flexplate under these conditions.

TASK A.1.3

5. A vehicle is being diagnosed for a transmission fluid leak near the bell housing area. Technician A says that a failed output shaft seal could cause a leak in this area. Technician B says that a leaking torque converter could cause a leak in this area. Who is correct?

   A. A only
   B. B only
   C. Both A and B
   D. Neither A nor B

   Answer A is incorrect. A failed output shaft seal would cause a transmission leak near the output shafts or drive shaft.

   Answer B is correct. Only Technician B is correct. A leaking torque converter could cause transmission fluid to be present in the bell housing area.

   Answer C is incorrect. Only Technician B is correct.

   Answer D is incorrect. Technician B is correct.

TASK A.1.3

6. All of the following problems could cause low transmission fluid level EXCEPT:

   A. Leaking transmission pan gasket
   B. Cracked valve body channel plate
   C. Faulty modulator valve
   D. Output shaft seal

   Answer A is incorrect. A leaking transmission pan gasket would cause the transmission fluid level to become low.

   Answer B is correct. A cracked valve body channel plate would likely cause shifting problems, but would not cause an external fluid leak.

   Answer C is incorrect. A faulty modulator that leaks transmission fluid through the vacuum hose could cause the transmission fluid level to drop.

   Answer D is incorrect. A leaking output shaft seal could cause the transmission fluid level to drop.

TASK A.1.4

7. Technician A says that the governor pressure should be approximately 30 psi at 15 mph. Technician B says that governor pressure should be near zero when the vehicle is parked. Who is correct?

   A. A only
   B. B only
   C. Both A and B
   D. Neither A nor B

   Answer A is incorrect. The governor pressure should be approximately 15 psi at 15 mph.

   Answer B is correct. Only Technician B is correct. The governor pressure is typically equal to the vehicle speed.

   Answer C is incorrect. Only Technician B is correct.

   Answer D is incorrect. Technician B is correct.

pressure to the valve body that is proportional to vehicle speed.

Answer B is incorrect. The manual valve has a great affect on the line pressure. The line pressure rises sharply when the gear selector is moved to reverse as well as manual low gear.

Answer C is incorrect. The throttle valve has a great affect on the line pressure. The line pressure increases as the throttle valve moves from low engine load to higher engine load.

Answer D is incorrect. The gear selector moves the manual valve. The manual valve has a great affect on the line pressure. The line pressure rises sharply when the gear selector is moved to reverse as well as manual low gear.

9. A vehicle has a stall speed that is 325 rpm below the standard specification. Technician A says that a locked torque converter stator could be the cause. Technician B says that a weak transmission oil pump could be the cause. Who is correct?

TASK A.1.5

   A. A only
   B. B only
   C. Both A and B
   D. Neither A nor B

Answer A is incorrect. A locked torque converter stator would not cause a decreased stall speed. This fault would impede vehicle power at highway speeds.

Answer B is incorrect. A weak transmission oil pump would cause the transmission clutch packs and bands to not hold well, which would cause higher than normal stall speed.

Answer C is incorrect. Neither Technician is correct.

Answer D is correct. Neither Technician is correct. Low stall speed could be caused by the stator freewheeling at heavy torque situations. Also, any problem associated with a weak engine could cause reduced stall speed.

10. All of the following transmission components are being tested during a stall test EXCEPT:

TASK A.1.5

   A. Oil pump capacity
   B. Band holding capacity
   C. Clutch pack holding capacity
   D. Torque converter stator

Answer A is correct. The oil pump capacity is not being tested during a stall test. A pressure test would need to be performed to determine the oil pump capacity.

Answer B is incorrect. The band holding capacity is tested during a stall test. If the band slips during a stall test, the stall speed will be higher than normal.

Answer C is incorrect. The clutch holding capacity is being tested during a stall test. If the clutch pack slips during a stall test, the stall speed will be higher than normal.

Answer D is incorrect. The torque converter stator is being tested during a stall test. If the stator one-way clutch does not hold, the stall speed will be lower than normal.

TASK A.1.6

11. Technician A says that the torque converter lockup clutch may be controlled by a lockup solenoid. Technician B says that the torque converter lockup clutch does not engage at speeds below about 55 miles per hour. Who is correct?

    A. A only

    B. B only

    C. Both A and B

    D. Neither A nor B

Answer A is correct. Only Technician A is correct. Many lockup torque converter clutches are controlled by a lockup solenoid that redirects fluid through the converter when the clutch is engaged.

Answer B is incorrect. Most lockup torque converter clutches will engage at speeds above approximately 35 miles per hour.

Answer C is incorrect. Only Technician A is correct.

Answer D is incorrect. Technician A is correct.

TASK A.1.6

12. Technician A says that the torque converter lockup clutch can be commanded to turn on by using a bi-directional scan tool. Technician B says that the engine RPM should increase approximately 150 rpm when the converter lockup clutch engages. Who is correct?

    A. A only

    B. B only

    C. Both A and B

    D. Neither A nor B

Answer A is correct. Only Technician A is correct. Many bi-directional scan tools have an output test to engage the torque converter clutch.

Answer B is incorrect. The engine RPM should decrease approximately 150 rpm when the torque converter lockup clutch engages.

Answer C is incorrect. Only Technician A is correct.

Answer D is incorrect. Technician A is correct.

TASK A.2.1

13. A vehicle with an electronic transmission is being diagnosed. Technician A says that it is wise to perform a line pressure test prior to conducting a road test. Technician B says that it is wise to check the transmission fluid prior to conducting a road test. Who is correct?

    A. A only

    B. B only

    C. Both A and B

    D. Neither A nor B

Answer A is incorrect. It would not be practical to perform a line pressure test prior to conducting a road test. The line test is performed on some vehicles later in the diagnosis process.

Answer B is correct. Only Technician B is correct. It is advisable to check the transmission fluid level prior to conducting a road test. Doing this will ensure the fluid level is not too low to drive the vehicle. In addition, the technician can uncover many problems by inspecting the color and smell of the transmission fluid.

Answer C is incorrect. Only Technician B is correct.

Answer D is incorrect. Technician B is correct.

Answer A is incorrect. The torque converter lockup clutch should be monitored during the road test to provide feedback about the functionality of that system.

Answer B is incorrect. The torque converter lockup clutch should be monitored during the road test to provide feedback about the functionality of that system.

Answer C is correct. The engine oil pressure is not among the items that are closely monitored when performing a road test for a transmission-related concern.

Answer D is incorrect. The smoothness of the up-shifts should be monitored during the road test to provide feedback about the shifting system.

15. The pressure control solenoid coil has an open circuit. Technician A says that the transmission will likely jerk when it is shifted from park to reverse. Technician B says that the transmission will likely still have smooth up-shifts. Who is correct?

TASK A.2.2

    A.   A only

    B.   B only

    C.   Both A and B

    D.   Neither A nor B

Answer A is correct. Only Technician A is correct. An open circuit in the pressure control solenoid will cause the line pressure to be the maximum level at all times. This high line pressure will cause all shifts of the transmission to be harsh.

Answer B is incorrect. The transmission will not have smooth up-shifts due to the line pressure being peaked out.

Answer C is incorrect. Only Technician A is correct.

Answer D is incorrect. Technician A is correct.

16. A vehicle with an electronic transmission has been tested for transmission pump output pressure. The pressure test results are below the specification in all ranges. All of the following conditions could cause this problem EXCEPT:

TASK A.2.2

    A.   Restricted transmission filter

    B.   Slipping clutch pack

    C.   Punctured transmission fluid pickup tube

    D.   Excess transmission pump clearance

Answer A is incorrect. A restricted transmission filter could cause reduced pressure in all ranges.

Answer B is correct. A slipping clutch pack would not cause low pump pressure. Low pump pressure would likely result in a slipping clutch pack.

Answer C is incorrect. A punctured transmission fluid pickup tube could cause reduced pressure in all ranges. This problem would likely cause clutch packs to slip in all ranges.

Answer D is incorrect. Excess transmission pump clearance would cause lower than normal transmission pressure in every range.

TASK A.2.3

17. All of the following conditions will cause the torque converter clutch to disengage EXCEPT:

    A. Applying the brakes
    B. Heavy throttle application
    C. Decreasing vehicle speed below minimum engagement specification
    D. Shifting the gear shift into overdrive

    Answer A is incorrect. Applying the brakes would cause the torque converter clutch to disengage due to the brake switch being a major input to the torque converter clutch.

    Answer B is incorrect. Heavy throttle would cause the torque converter clutch to disengage due to the high engine load.

    Answer C is incorrect. Dropping below the minimum vehicle speed setting would cause the torque converter clutch to disengage. Many manufacturers have a minimum speed of approximately 35 miles per hour.

    Answer D is correct. Shifting the gear shift to overdrive would not cause the torque converter to disengage. In fact, the torque converter clutch is most effective in the top gear and helps improve fuel economy by reducing the engine RPM at cruising speed.

TASK A.2.3

18. A vehicle is being diagnosed for a problem with the torque converter clutch system. Technician A says that the torque converter clutch should be engaged at 45 mph after the engine has warmed up. Technician B says that the torque converter clutch may disengage when the vehicle climbs a steep hill at highway speed. Who is correct?

    A. A only
    B. B only
    C. Both A and B
    D. Neither A nor B

    Answer A is incorrect. Technician B is also correct.

    Answer B is incorrect. Technician A is also correct.

    Answer C is correct. Both Technicians are correct. The torque converter clutch system should be engaged by the time the vehicle reaches 45 mile per hour with a warmed-up engine. In addition, it is common for the torque converter clutch to disengage when traveling up steep hills at highway speeds.

    Answer D is incorrect. Both Technicians are correct.

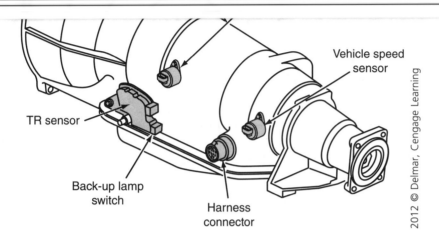

Vehicle speed sensor

TR sensor

Back-up lamp switch

Harness connector

2012 © Delmar, Cengage Learning

19. Referring to the figure above, Technician A says that the TR sensor is used to signal the selected range to the PCM. Technician B says that the harness connector can be used to test the resistance of the internal shift solenoids with an ohmmeter. Who is correct?

TASK A.2.4

A. A only

B. B only

C. Both A and B

D. Neither A nor B

Answer A is incorrect. Technician B is also correct.

Answer B is incorrect. Technician A is also correct.

Answer C is correct. Both Technicians are correct. The TR sensor is the transmission component that signals the selected range to the PCM or TCM. The transmission harness connector is a common location that is used to test the resistance of the internal shift solenoids. Testing at this location will make it unnecessary to remove the transmission pan to test the electrical circuits inside the transmission.

Answer D is incorrect. Both Technicians are correct.

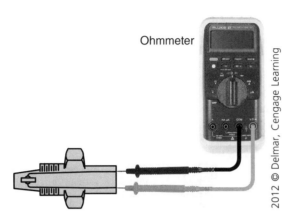

Ohmmeter

TASK A.2.4

20. Referring to the figure above, Technician A says that the fluid temperature sensor resistance should decrease as the temperature of the sensor increases. Technician B says that the fluid temperature sensor resistance should increase as the temperature decreases. Who is correct?

   A. A only

   B. B only

   C. Both A and B

   D. Neither A nor B

Answer A is incorrect. Technician B is also correct.

Answer B is incorrect. Technician A is also correct.

Answer C is correct. Both Technicians are correct. Most thermistors that are used as temperature sensors on cars and trucks are of the negative temperature coefficient (NTC) design, which causes the resistance and the temperature of the device to be inversely related.

Answer D is incorrect. Both Technicians are correct.

TASK A.2.5

21. The battery housing received some damage from driving a vehicle on rough roads. Electrolyte spilled all over the battery tray. Technician A says that brake cleaner should be used to clean the area. Technician B says that baking soda could be used to neutralize the battery acid. Who is correct?

   A. A only

   B. B only

   C. Both A and B

   D. Neither A nor B

Answer A is incorrect. Brake cleaner should never be used around spilled electrolyte.

Answer B is correct. Only Technician B is correct. Baking soda can be used to neutralize the spilled electrolyte acid of a battery.

Answer C is incorrect. Only Technician B is correct.

Answer D is incorrect. Technician B is correct.

not likely cause charging system problems.

Answer B is incorrect. A loose terminal connection at the charging system "Maxi Fuse" would cause voltage drop in the charging circuit, which would reduce the output voltage.

Answer C is incorrect. A loose generator mounting bracket would cause voltage drop in the ground side of the charging circuit, which would reduce the output voltage.

Answer D is incorrect. A damaged charging output wire would cause voltage drop in the charging circuit, which would reduce the output voltage.

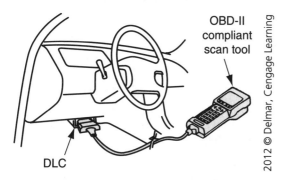

OBD-II compliant scan tool

DLC

2012 © Delmar, Cengage Learning

23. The tool shown in the picture above can be used for which of the following purposes?

TASK A.2.6

A. Testing the vehicle battery

B. Checking the charging system output

C. Accessing the transmission computer data list

D. Checking for the supply voltage (B+) at the transmission computer

Answer A is incorrect. A digital battery tester would be used to test the vehicle battery. The item in the picture is a scan tool.

Answer B is incorrect. A digital meter would be used to check the charging system output capabilities.

Answer C is correct. The scan tool in the picture would be used to access the transmission computer data list. The tool can also be used to retrieve trouble codes and perform output tests.

Answer D is incorrect. A digital meter would be used to check the supply voltage (B+) at the transmission computer.

TASK A.2.6

24. A vehicle has low power when taking off from a stop, but the power seems normal at highway speeds. Technician A says that the engine exhaust system could be restricted. Technician B says that the torque converter stator could be defective. Who is correct?

    A. A only

    B. B only

    C. Both A and B

    D. Neither A nor B

    Answer A is incorrect. A restricted exhaust system would cause low power at all times.

    Answer B is correct. Only Technician B is correct. A defective torque converter stator could cause low vehicle power at low speeds but would not be evident at highway speeds.

    Answer C is incorrect. Only Technician B is correct.

    Answer D is incorrect. Technician B is correct.

TASK A.2.7

25. A problem with which of the following electronic transmission components would be most likely to cause the transmission to have extremely firm shifts?

    A. Range sensor

    B. Output speed sensor

    C. Input shaft speed sensor

    D. Pressure control solenoid

    Answer A is incorrect. A range sensor fault could cause a transmission to not have all of its ranges, but it would not likely cause very firm shifts.

    Answer B is incorrect. A fault with the output speed sensor could cause the transmission to enter "limp mode," but would not cause very harsh shifts.

    Answer C is incorrect. A fault with the input speed sensor could cause the transmission to enter "limp mode," but would not cause very harsh shifts.

    Answer D is correct. A problem with the pressure control solenoid could cause the transmission to have very firm shifts due to increased transmission pressures.

TASK A.2.7

26. The main power relay for the electronic control system has failed. Technician A says that the vehicle will not move in forward or reverse. Technician B says that this fault should set a diagnostic trouble code (DTC) in the transmission computer. Who is correct?

    A. A only

    B. B only

    C. Both A and B

    D. Neither A nor B

    Answer A is incorrect. Most electronic transmissions and transaxles will still move forward and reverse if the main power is lost at the transmission or transaxle.

    Answer B is correct. Only Technician B is correct. A diagnostic trouble code (DTC) should be recorded in the transmission computer when any major electronic system malfunctions within the transmission control system.

    Answer C is incorrect. Only Technician B is correct.

    Answer D is incorrect. Technician B is correct.

D.   Neither A nor B

Answer A is incorrect. Technician B is also correct.

Answer B is incorrect. Technician A is also correct.

Answer C is correct. Both Technicians are correct. A misadjusted range sensor or park/neutral switch can sometimes cause a no-crank condition. Typically, a vehicle will have one or the other of these components. The adjustment of these components is vital to the engine starting correctly, as well as the back-up lights working as they should.

Answer D is incorrect. Both Technicians are correct.

28.   A transmission will move in forward and reverse but has no up-shifts. Technician A says that a broken kick-down cable could be the cause. Technician B says that a faulty governor valve could be the cause. Who is correct?

TASK B.2

A.   A only

B.   B only

C.   Both A and B

D.   Neither A nor B

Answer A is incorrect. A broken kick-down cable could cause soft and early shifts, but the transmission would still be able to up-shift.

Answer B is correct. Only Technician B is correct. A faulty governor could cause the transmission to not have any up-shifts. The governor sends a hydraulic signal to the valve body that is proportional to vehicle speed.

Answer C is incorrect. Only Technician B is correct.

Answer D is incorrect. Technician B is correct.

29.   A transmission needs to have an extension housing gasket replaced. Technician A says that the drive shaft will need to be removed during this process. Technician B says that the transmission will have to be removed from the vehicle to perform this repair on most transmissions. Who is correct?

TASK B.3

A.   A only

B.   B only

C.   Both A and B

D.   Neither A nor B

Answer A is correct. Only Technician A is correct. The drive shaft has to be removed to replace the extension housing gasket.

Answer B is incorrect. The extension housing gasket can be serviced without removing the transmission on most transmissions.

Answer C is incorrect. Only Technician A is correct.

Answer D is incorrect. Technician A is correct.

TASK B.4

30. Technician A says that the transmission vent should be checked when a repetitive seal failure occurs. Technician B says that the drive shaft has to be removed to replace the extension housing output seal. Who is correct?

   A. A only
   B. B only
   C. Both A and B
   D. Neither A nor B

   Answer A is incorrect. Technician B is also correct.

   Answer B is incorrect. Technician A is also correct.

   Answer C is correct. Both Technicians are correct. A clogged transmission vent can allow pressure to build up inside the transmission and cause repeated seal failures. It is common to remove the drive shaft when replacing the extension housing output seal.

   Answer D is incorrect. Both Technicians are correct.

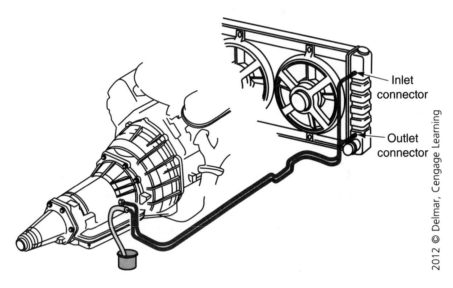

Inlet connector

Outlet connector

2012 © Delmar, Cengage Learning

TASK B.5

31. Referring to the figure above, Technician A says that the cooler flow test should be performed at each transmission service. Technician B says that the radiator may have to be replaced if the amount measured is less than one quart in 20 seconds during the test. Who is correct?

   A. A only
   B. B only
   C. Both A and B
   D. Neither A nor B

   Answer A is incorrect. It is not common practice to test the flow rate of the transmission cooler at each transmission service. This test should be performed whenever there is evidence of overheated transmission fluid or at each rebuild procedure.

   Answer B is correct. Only Technician B is correct. The transmission fluid cooler flow test should produce about one quart per 20 seconds. If the test results are less than this, the cooler should be flushed or replaced.

   Answer C is incorrect. Only Technician B is correct.

   Answer D is incorrect. Technician B is correct.

D.   Neither A nor B

Answer A is incorrect. The torque-to-yield method of tightening fasteners would not be used on the valve body bolts. This method of tightening is typically used on fasteners with a high clamp load.

Answer B is correct. Only Technician B is correct. It is advisable to mark the valve body bolts as they are removed in order to reinstall them into the correct location.

Answer C is incorrect. Only Technician B is correct.

Answer D is incorrect. Technician B is correct.

33.   A transmission has a broken 1-2 accumulator spring. Technician A says that the transmission will likely slip during the first-to-second-gear shift sequence. Technician B says that the line pressure will likely increase due to the broken spring. Who is correct?

TASK B.8

A.   A only
B.   B only
C.   Both A and B
D.   Neither A nor B

Answer A is incorrect. A broken 1-2 accumulator spring would cause a harsh shift from first to second gear.

Answer B is incorrect. A broken 1-2 accumulator spring would not affect the line pressure in the transmission. It would only cause a harsh shift from first to second.

Answer C is incorrect. Neither Technician is correct.

Answer D is correct. Neither Technician is correct. A broken 1-2 accumulator spring would cause a harsh shift from first to second gear.

34.   Which of the following tests can be performed with a fused jumper wire?

TASK B.9

A.   By-passing a rheostat
B.   By-passing a thermistor
C.   By-passing the load side of a relay
D.   By-passing a potentiometer

Answer A is incorrect. A jumper wire is not typically used to by-pass a rheostat because of the possibility of a current overload in the circuit.

Answer B is incorrect. A jumper wire is not typically used to by-pass a thermistor because of the possibility of a current overload in the circuit.

Answer C is correct. It is acceptable to use a fused jumper wire to by-pass the load side of a relay. This test gives feedback to the technician about whether the load side of the circuit is functioning.

Answer D is incorrect. A jumper wire is not typically used to by-pass a potentiometer because of the possibility of a current overload in the circuit.

TASK B.9

35. Technician A says that a test light can be used to test for voltage on computer data circuits. Technician B says that a test light can be used to check for power on the input and output sides of fuses at the fuse panel. Who is correct?

A. A only

B. B only

C. Both A and B

D. Neither A nor B

Answer A is incorrect. A test light should never be used to test any type of computer circuit. A high-impedance meter is necessary to safely test these circuits.

Answer B is correct. Only Technician B is correct. Using a test light to test fuses at the fuse panel or power distribution center is acceptable. A good fuse will create light on the input and output side of a fuse.

Answer C is incorrect. Only Technician B is correct.

Answer D is incorrect. Technician B is correct.

TASK B.9

36. Which of the following definitions best describes a relay?

A. A magnetic switch

B. A metallic component that is used to tie electrical circuits together

C. A plastic housing that is used to hold terminals, as well as to plug into electrical items

D. A joint where two or more electrical circuits connect to each other

Answer A is correct. A relay is often described as a magnetic switch because it uses a magnetic coil to close a set of contacts. Relays are used to power up items that draw a high level of current flow.

Answer B is incorrect. A terminal is described as a metallic component that is used to tie an electrical circuit together.

Answer C is incorrect. A connector is described as a plastic housing that is used to hold terminals, as well as to plug into electrical items.

Answer D is incorrect. A splice is described as a joint where two or more electrical circuits connect to each other.

TASK B.10

37. A vehicle is being diagnosed for a problem of the accelerator pedal stuck in the wide-open throttle position. Which of the following components could cause this condition?

A. Broken cruise control cable

B. Broken throttle cable

C. Broken engine mount

D. Shorted throttle position sensor

Answer A is incorrect. A broken cruise control cable would cause the cruise control system to be inoperative.

Answer B is incorrect. A broken throttle cable would cause the accelerator pedal to be inoperative.

Answer C is correct. A broken engine mount could cause the engine to shift enough to stretch the accelerator cable and make the throttle hang wide open.

Answer D is incorrect. A shorted throttle positions sensor would cause a "service engine" indicator to illuminate, but it would not cause the accelerator pedal to stick in the wide-open throttle position.

gear train.

Answer B is correct. A common location for the transmission filter is inside the transmission pan assembly.

Answer C is incorrect. The transmission filter would not be located inside the transmission cooler.

Answer D is incorrect. The transmission filter would not be located inside the torque converter.

39. A vehicle has a fluid leak around the bell housing area. Which of the following faults would be LEAST LIKELY to cause this problem?

TASK C.1.1

    A. Transmission front pump seal

    B. Transmission bottom pan gasket

    C. Engine rear main seal

    D. Transmission front pump retaining bolts

Answer A is incorrect. A leaking transmission front pump seal would cause fluid to be present in the bell housing area.

Answer B is correct. A leaking transmission pan gasket would not cause fluid to be present in the bell housing area.

Answer C is incorrect. A leaking engine rear main seal would cause fluid (oil) to be present in the bell housing area. The technician should always attempt to identify the type of fluid that is leaking in order to determine its source.

Answer D is incorrect. Leaking transmission front pump retaining bolts would cause fluid to be present in the bell housing area. A technician should always apply thread sealer to the threads of the front pump retaining bolts to prevent this type of leak from occurring.

40. A transmission is ready for installation. Which of the following inspections would be LEAST LIKELY to be needed prior to this installation?

TASK C.1.2, C.1.1

    A. Crankshaft pilot bore

    B. Cracks in flexplate

    C. Clutch pack clearance

    D. Elongated mounting holes in flexplate

Answer A is incorrect. The crankshaft pilot bore should be inspected for scrapes and burrs prior to installing the transmission.

Answer B is incorrect. The flexplate should be closely inspected for cracks prior to installing the transmission.

Answer C is correct. The clutch pack clearance would not be inspected at this point. Clutch pack clearance should be tested as the transmission is being assembled.

Answer D is incorrect. The flexplate fastener mounting holes should be closely inspected for elongation prior to installing the transmission.

TASK C.1.3

41. Technician A says that the transmission cooler should be flushed every time that the transmission fluid is serviced. Technician B says that the cooler should be reverse flushed if possible to increase the chances of a thorough cleaning. Who is correct?

    A. A only

    B. B only

    C. Both A and B

    D. Neither A nor B

Answer A is incorrect. It is not necessary to flush the transmission cooler at every transmission fluid service. It is recommended to flush the transmission fluid at each rebuild or transmission replacement.

Answer B is correct. Only Technician B is correct. Reverse flushing a transmission cooler increases the likelihood of removing dirt and foreign particles from the cooler assembly.

Answer C is incorrect. Only Technician B is correct.

Answer D is incorrect. Technician B is correct.

TASK C.1.3

42. Which component has to be replaced in order to replace the primary transmission fluid cooler?

    A. Condenser

    B. Reserve tank

    C. Radiator

    D. Water pump

Answer A is incorrect. The condenser is the heat exchanger for the air conditioning system that is mounted in front of the radiator.

Answer B is incorrect. The reserve tank is part of the engine cooling system. This device holds extra engine coolant as well as allowing the hot coolant to be stored until the engine cools down.

Answer C is correct. The primary transmission cooler is located in the radiator tank. The radiator can either be replaced or serviced at a radiator shop for the cooler replacement.

Answer D is incorrect. The water pump is the device that forces engine coolant to be moved throughout the engine, the heater system, and the radiator.

TASK C.2.3

43. Which of the following statements about bearing preload is LEAST LIKELY to be correct?

    A. A bearing preload that is set too tight will loosen up once the unit heats up.

    B. A torque wrench is needed to set the bearing preload.

    C. Bearing preload is sometimes checked by testing the turning effort of a component.

    D. Bearings must be lubricated in order to perform under an extended load.

Answer A is correct. Bearing preload that is set too tight will likely get tighter as the unit heats up.

Answer B is incorrect. A torque wrench is used to tighten up fasteners when setting the bearing preload.

Answer C is incorrect. Bearing preload is checked on many transaxles by testing the turning effort on the gears. A dial-type or deflecting beam-type torque wrench is needed to perform this test.

Answer D is incorrect. All bearings need to be lubricated with some type of lubricant when they are operating under a load.

D.   Neither A nor B

Answer A is incorrect. A shop rag should not be used to clean an internal oil passage because of the likelihood of leaving a piece of lint in the transmission shaft.

Answer B is correct. Only Technician B is correct. Solvent is commonly used to internally clean internal oil passages on transmission components. Compressed air can be used to dry out the various components that are cleaned with solvent.

Answer C is incorrect. Only Technician B is correct.

Answer D is incorrect. Technician B is correct.

45.   Which of the following actions would be LEAST LIKELY performed when inspecting a planetary gear set?

TASK C.2.8

A.   Measure the pinion gear end-play with a feeler gauge.

B.   Measure the bearing preload of the pinion gears.

C.   Inspect the pinion gears for damaged teeth.

D.   Measure the pinion gear end-play with dial indicator.

Answer A is incorrect. A feeler gauge is often used to measure pinion gear end-play on planetary carriers.

Answer B is correct. It is not a common practice to measure the bearing preload on pinion gears of planetary gears.

Answer C is incorrect. It is important to check the pinion gears for damaged teeth to make sure the gears can be reused.

Answer D is incorrect. A dial indicator can often be used to measure pinion gear end-play on planetary carriers.

46.   A mating surface for the valve body is found to be warped. Technician A says that the transmission will have to be machined in order to correct the warpage. Technician B says that a special spacer plate may be used to accommodate the uneven surface. Who is correct?

TASK C.2.9

A.   A only

B.   B only

C.   Both A and B

D.   Neither A nor B

Answer A is incorrect. It is not common to machine transmission cases to remove warped surfaces. The transmission case is typically replaced when excess warpage is found.

Answer B is incorrect. A spacer plate will not correct a warped transmission case.

Answer C is incorrect. Neither Technician is correct.

Answer D is correct. Neither Technician is correct. The transmission case would need to be replaced when excess warpage is found on the mating surface for the valve body.

TASK C.2.11

47. Which of the following tools would most likely be used when measuring the backlash on the final drive gears in a transaxle?

    A.   Dial indicator

    B.   Slide caliper

    C.   Outside micrometer

    D.   Feeler gauge

Answer A is correct. A dial indicator is the normal tool that is used when very small amounts of movement need to be measured.

Answer B is incorrect. A slide caliper can be used for measuring outside diameter and inside diameter, as well as depth.

Answer C is incorrect. An outside micrometer is used to measure outside diameter on some transmission parts.

Answer D is incorrect. A feeler gauge is used to measure very small clearances between two parts. A feeler gauge is commonly used to measure clutch pack clearance.

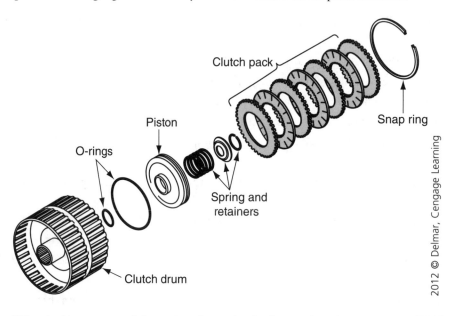

48. What is the purpose of the spring shown in the figure above?

TASK C.3.1

    A.   To apply the piston when fluid stops flowing into the clutch pack

    B.   To release the piston when fluid stops flowing into clutch pack

    C.   To hold pressure on the clutch pack when fluid is sent to the assembly

    D.   To release pressure on the clutch pack when fluid is sent to the assembly

Answer A is incorrect. The piston is applied when fluid flows into the clutch pack.

Answer B is correct. The return spring in the picture is used to release the piston when transmission fluid stops flowing into the clutch pack.

Answer C is incorrect. The pressure of the spring is overcome while the clutch pack is receiving transmission fluid.

Answer D is incorrect. The clutch pack is applied when fluid pressure is sent to the clutch pack.

Pressure
plate
2012 © Delmar

49. Which of the follow procedures is most likely being performed in the figure above?

    A. Clutch pack apply plate selection

    B. Clutch pack clearance test

    C. Clutch pack snap ring removal

    D. Clutch pack piston return spring service

TASK C.3.2

Answer A is incorrect. A micrometer would be involved in selecting the apply plate for the clutch pack.

Answer B is correct. The clutch pack clearance is being checked by sliding a feeler gauge between the snap ring and the pressure plate.

Answer C is incorrect. A screwdriver is typically used to remove the clutch pack snap ring.

Answer D is incorrect. A spring compressor is typically used to service the clutch pack piston return spring.

50. Technician A says that air testing a servo assembly will reveal a faulty servo piston seal. Technician B says that air testing a servo assembly will reveal a faulty servo piston return spring. Who is correct?

    A. A only

    B. B only

    C. Both A and B

    D. Neither A nor B

TASK C.3.2

Answer A is incorrect. Technician B is also correct.

Answer B is incorrect. Technician A is also correct.

Answer C is correct. Both Technicians are correct. Performing an air test on a servo will provide feedback about the servo piston seal and the servo piston return spring. A good piston and return spring will produce a popping sound when air is applied to the servo and will make a noticeable thud sound when the air is removed from the servo circuit.

Answer D is incorrect. Both Technicians are correct.

PREPARATION EXAM 5—ANSWER KEY

1.	B	**21.**	A	**41.**	A
2.	B	**22.**	C	**42.**	B
3.	B	**23.**	A	**43.**	B
4.	C	**24.**	C	**44.**	C
5.	C	**25.**	D	**45.**	A
6.	A	**26.**	C	**46.**	B
7.	D	**27.**	A	**47.**	B
8.	B	**28.**	C	**48.**	A
9.	C	**29.**	D	**49.**	C
10.	C	**30.**	A	**50.**	A
11.	D	**31.**	B		
12.	C	**32.**	C		
13.	C	**33.**	B		
14.	A	**34.**	C		
15.	C	**35.**	C		
16.	B	**36.**	B		
17.	D	**37.**	C		
18.	C	**38.**	B		
19.	A	**39.**	B		
20.	B	**40.**	B		

PREPARATION EXAM 5—EXPLANATIONS

TASK A.1.1

1. During a road test, the technician notices that the transmission slips when making a sharp turn. Technician A says that the pressure control solenoid could be sticking. Technician B says that the transmission fluid level could be low. Who is correct?

   A.   A only

   B.   B only

   C.   Both A and B

   D.   Neither A nor B

   Answer A is incorrect. A sticking pressure control solenoid is not likely to cause reduced pressure while making turns.

   Answer B is correct. Only Technician B is correct. Low transmission fluid could cause the transmission to slip when making a turn due to the fluid level dropping below the pickup screen on the filter.

   Answer C is incorrect. Only Technician B is correct.

   Answer D is incorrect. Technician B is correct.

Answer A is incorrect. A misadjusted throttle valve cable would cause early or late shift points.

Answer B is correct. A ruptured vacuum modulator could cause transmission fluid to be pulled into the engine through the vacuum hose that connects to the modulator.

Answer C is incorrect. A restricted transmission cooler would cause a hard part failure due to overheating.

Answer D is incorrect. A faulty MAP sensor could cause a service engine light and a likely engine performance problem.

3. Which of the following faults would be most likely to cause a clicking noise that increases when the transmission is shifted into drive and goes away when the transmission is shifted into park?

TASK A.1.2

   A. Worn transmission pump
   B. Loose torque converter bolts
   C. Slipping transmission clutch pack
   D. Torque converter lockup clutch slipping

Answer A is incorrect. A worn transmission pump would cause a whining sound that increases with engine speed.

Answer B is correct. Loose torque converter bolts could cause a noise that is worse when the transmission is engaged into gear.

Answer C is incorrect. A slipping transmission clutch pack would cause the engine RPM to rise without an increase in torque being transmitted.

Answer D is incorrect. A slipping torque converter lockup clutch would cause a shudder feeling as the clutch was being engaged.

4. A vehicle is being diagnosed for a power train vibration at 50 to 60 mph. Technician A says that a bent drive shaft could be the cause. Technician B says that a faulty bent pinion yoke could be the problem. Who is correct?

TASK A.1.2

   A. A only
   B. B only
   C. Both A and B
   D. Neither A nor B

Answer A is incorrect. Technician B is also correct.

Answer B is incorrect. Technician A is also correct.

Answer C is correct. Both Technicians are correct. A bent drive shaft as well as a bent pinion yoke could cause a vibration at 50 to 60 mph. The drive shaft would need to be removed and closely inspected to determine which fault was present.

Answer D is incorrect. Both Technicians are correct.

TASK A.1.3

5. Technician A says that the engine should be running when checking the transmission fluid. Technician B says that the dipstick should reveal which gear to put the transmission in to check the transmission fluid. Who is correct?

   A. A only

   B. B only

   C. Both A and B

   D. Neither A nor B

   Answer A is incorrect. Technician B is also correct.

   Answer B is incorrect. Technician A is also correct.

   Answer C is correct. Both Technicians are correct. The engine should be running when checking the transmission fluid. It is common to see the instructions on the transmission dipstick on which gear to put the transmission into when checking the fluid.

   Answer D is incorrect. Both Technicians are correct.

TASK A.1.4

6. A transmission pressure test has been performed on a vehicle. Which of the following pieces of data would be LEAST LIKELY revealed from this test?

   A. Clutch return spring pressure

   B. Servo application pressure

   C. Mainline pressure

   D. Governor pressure

   Answer A is correct. A pressure test will not test the clutch return spring pressure. This data would be measured with a spring rate test device.

   Answer B is incorrect. The servo application pressure could be determined by checking the pressure in the servo circuit.

   Answer C is incorrect. The mainline pressure could be determined by checking the pressure in a mainline circuit.

   Answer D is incorrect. The governor pressure could be determined by checking the pressure in a governor circuit.

TASK A.1.4

7. A vehicle will back-up and go into first gear without a problem. The vehicle will not up-shift until very high engine RPM. Which of the following conditions would most likely cause the problem?

   A. Manual valve is out of adjustment.

   B. Transmission fluid is overfilled.

   C. Band is out of adjustment.

   D. Governor valve is sticking.

   Answer A is incorrect. A misadjusted manual valve would cause the transmission to not have the correct ranges.

   Answer B is incorrect. An overfilled transmission would cause the fluid to aerate and slip while driving.

   Answer C is incorrect. A misadjusted band would cause the transmission to slip if it were adjusted too loose and would cause the transmission to take off in the wrong gear if adjusted too tight.

   Answer D is correct. A sticking governor valve would have lower than normal output pressure, which would cause late up-shifts.

D.   Neither A nor B

Answer A is incorrect. A locked torque converter stator would cause the vehicle to have normal power while taking off but lack power at highway speeds.

Answer B is correct. Only Technician B is correct. A stall speed within 25 rpm of the specification is considered normal.

Answer C is incorrect. Only Technician B is correct.

Answer D is incorrect. Technician B is correct.

9. Technician A says that the stall test should not be continued for longer than five seconds. Technician B says that performing stall tests increases the temperature of the transmission fluid greatly. Who is correct?

TASK A.1.5

A.   A only

B.   B only

C.   Both A and B

D.   Neither A nor B

Answer A is incorrect. Technician B is also correct.

Answer B is incorrect. Technician A is also correct.

Answer C is correct. Both Technicians are correct. Performing stall tests on a transmission quickly increases the temperature of the fluid so the test should be limited to about five seconds at a time. In addition, the technician should allow the engine to idle for a couple of minutes between stall tests to allow the fluid to be cooled down.

Answer D is incorrect. Both Technicians are correct.

10. Technician A says that some transmission computers will engage the torque converter lockup clutch under heavy loads when the transmission fluid temperature rises above safe levels. Technician B says that the torque converter lockup clutch is locked up when the pump and turbine are near the same RPM. Who is correct?

TASK A.1.6

A.   A only

B.   B only

C.   Both A and B

D.   Neither A nor B

Answer A is incorrect. Technician B is also correct.

Answer B is incorrect. Technician A is also correct.

Answer C is correct. Both Technicians are correct. It is a common strategy for the transmission computer to engage the torque converter lockup clutch when the fluid temperature rises too high. This prevents further heating of the fluid by locking the pump and the turbine together. In normal conditions, the torque converter lockup clutch receives the signal to engage when the pump and turbine are near the same speed.

Answer D is incorrect. Both Technicians are correct.

TASK A.1.6

11. Technician A says that the flow to the transmission cooler is restricted when the torque converter lockup clutch is disengaged. Technician B says that the torque converter builds high levels of heat when the lockup clutch is engaged. Who is correct?

    A. A only

    B. B only

    C. Both A and B

    D. Neither A nor B

Answer A is incorrect. The flow of fluid to the transmission cooler is restricted when the torque converter lockup clutch is engaged.

Answer B is incorrect. The torque converter stops building heat when the torque converter lockup clutch is engaged.

Answer C is incorrect. Neither Technician is correct.

Answer D is correct. Neither Technician is correct. The torque converter is the main source of heat in the transmission. The fluid is cooled when it is pumped to the transmission cooler. When the lockup clutch is engaged, the torque converter stops building heat because the pump and turbine are turning at the same speed. In addition, the flow of fluid to the cooler is typically stopped when the torque converter lockup clutch is engaged.

TASK A.2.1

12. A vehicle with an electronic transmission is being diagnosed. Technician A says that a scan tool can expose possible problems during the road test. Technician B says that it is a good idea to check the transmission fluid prior to conducting a road test. Who is correct?

    A. A only

    B. B only

    C. Both A and B

    D. Neither A nor B

Answer A is incorrect. Technician B is also correct.

Answer B is incorrect. Technician A is also correct.

Answer C is correct. Both Technicians are correct. A scan tool can access transmission data and trouble codes as well as commanding output tests on an electronic transmission. It is a good practice to check the transmission fluid prior to conducting a road test to make sure the fluid level is high enough to safely drive the vehicle.

Answer D is incorrect. Both Technicians are correct.

TASK A.2.1

13. Which of the following items should be monitored during the road test of a vehicle with an electronic transmission?

    A. Transmission fluid color

    B. Transmission cooler flow rate

    C. Drive train noises

    D. Transmission fluid level

Answer A is incorrect. It would not be possible to monitor the color of the transmission color while driving a vehicle.

Answer B is incorrect. It would not be possible to monitor the transmission cooler flow rate during a road test.

Answer C is correct. It is always a good practice to monitor the engine and transmission noises while performing a road test. Unusual noises should be investigated when the vehicle is back at the shop.

Answer D is incorrect. It would not be possible to monitor the transmission fluid level during a road test.

Answer A is correct. Disconnecting the transmission fluid temperature (TFT) sensor would not cause the transmission line pressure to rise.

Answer B is incorrect. Most scan tools have the capacity to command the pressure to be increased by using the output test function.

Answer C is incorrect. Disconnecting the pressure control solenoid would cause the transmission line pressure to rapidly increase. This feature is a common trait for electronic transmissions that protects the unit if power is lost to the transmission.

Answer D is incorrect. Shifting the transmission to reverse while moving the throttle to wide open would cause the transmission pressure to rapidly increase.

15. A vehicle with an electronic pressure control solenoid is being diagnosed. Technician A says that that maximum pressure can be attained by putting the transmission in reverse and moving the throttle to wide-open throttle (WOT). Technician B says that maximum pressure can be attained by using a scan tool and commanding the pressure control solenoid to receive minimum amperage. Who is correct?

TASK A.2.2

A. A only
B. B only
C. Both A and B
D. Neither A nor B

Answer A is incorrect. Technician B is also correct.

Answer B is incorrect. Technician A is also correct.

Answer C is correct. Both Technicians are correct. Maximum line pressure can be achieved by shifting the transmission to reverse and accelerating to wide-open throttle. Maximum line pressure can also be achieved with a scan tool and commanding the pressure control solenoid to receive minimum amperage. This low amperage signal would cause the line pressure to rise to the maximum level.

Answer D is incorrect. Both Technicians are correct.

16. Which of the following components could cause the torque converter clutch and the cruise control to be inoperative?

TASK A.2.3

A. Cruise control switch
B. Brake switch
C. Overdrive switch
D. Headlight switch

Answer A is incorrect. The cruise control switch would not have an effect on the torque converter clutch.

Answer B is correct. The brake switch contributes to the operation of the cruise control and the torque converter clutch. If the brake switch fails, both of these systems could be inoperative.

Answer C is incorrect. The overdrive switch would not have an effect on the torque converter or the cruise control operation.

Answer D is incorrect. The headlight switch would not have an effect on the torque converter or the cruise control operation.

17. Which of the following conditions would be LEAST LIKELY to cause a drop in fuel economy?

   A. Coolant temperature sensor shorted out
   B. Torque converter clutch inoperative
   C. Engine thermostat stuck open
   D. Worn transmission oil pump

   Answer A is incorrect. A shorted coolant temperature sensor would have a great impact on the fuel economy for a vehicle. The engine control system would always run in open loop, which would increase fuel consumption.

   Answer B is incorrect. An inoperative torque converter clutch would cause a drop in fuel economy because the engine would have to run at a higher RPM at cruising speeds.

   Answer C is incorrect. A stuck open thermostat would cause a drop in fuel economy because the engine would take longer to warm up.

   Answer D is correct. A worn transmission pump would have no impact on vehicle fuel economy. A worn transmission pump may cause increased noise as well as poor transmission operation.

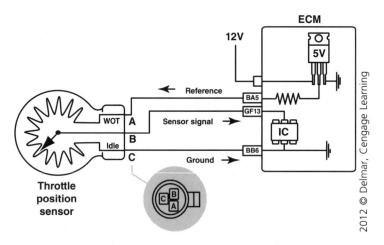

18. Referring to the figure above, which terminal would need to be back-probed to test the signal voltage for the throttle position sensor?

   A. Terminal A of the throttle position sensor
   B. Terminal BA5 of the ECM
   C. Terminal B of the throttle position sensor
   D. Terminal C of the throttle position sensor

   Answer A is incorrect. Terminal A of the throttle position sensor is the reference circuit for the sensor.

   Answer B is incorrect. Terminal BA5 of the ECM is the reference circuit for the sensor.

   Answer C is correct. Terminal B of the throttle position sensor is the signal circuit, which will change as the throttle angle is changed.

   Answer D is incorrect. Terminal C of the throttle position sensor is the ground circuit for the sensor.

databases.

Answer A is correct. Technical service bulletins do not provide free repairs for the customer if the vehicle is beyond the warranty coverage.

Answer B is incorrect. The vehicle manufacturer is the source for most technical service bulletins.

Answer C is incorrect. Technical service bulletins are issued to assist technicians in repairing pattern failures. The normal diagnostic routine should be followed prior to assure that the basic inspections are performed.

Answer D is incorrect. Aftermarket technicians have the ability to access technical service bulletins through professional databases.

20. A late-model vehicle is being diagnosed for a charging problem. The generator only charges at 12.4 volts. A voltage drop test is performed on the charging output wire and 1.8 volts is measured. Technician A says that a blown fusible link in the output circuit could be the cause. Technician B says that a loose nut at the charging output connector could be the cause. Who is correct?

TASK A.2.5

    A.   A only

    B.   B only

    C.   Both A and B

    D.   Neither A nor B

Answer A is incorrect. A blown fusible link would have produced a higher voltage drop.

Answer B is correct. Only Technician B is correct. A loose nut at the charging output connector could cause the excessive voltage drop in the charging circuit.

Answer C is incorrect. Only Technician B is correct.

Answer D is incorrect. Technician B is correct.

21. A maintenance-free battery is low on electrolyte. Technician A says a defective voltage regulator may cause this problem. Technician B says a loose alternator belt may cause this problem. Who is correct?

TASK A.2.5

    A.   A only

    B.   B only

    C.   Both A and B

    D.   Neither A nor B

Answer A is correct. Only Technician A is correct. A defective voltage regulator can cause overcharging and possible battery boil over. Voltage regulators can be located in the alternator or in the engine control module, depending on the application.

Answer B is incorrect. A loose alternator belt might cause undercharging, but not overcharging, which is what a low electrolyte level indicates.

Answer C is incorrect. Only Technician A is correct.

Answer D is incorrect. Technician A is correct.

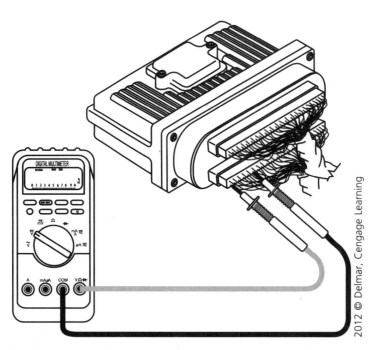

TASK A.2.6

22. Referring to the figure above, Technician A says that the meter leads are back-probing the module connector. Technician B says that all meters used in this manner need to be a high-impedance design. Who is correct?

A. A only

B. B only

C. Both A and B

D. Neither A nor B

Answer A is incorrect. Technician B is also correct.

Answer B is incorrect. Technician A is also correct.

Answer C is correct. Both Technicians are correct. The figure shows the meter leads being back-probed into the connectors. This method allows the circuit to function while the technician is performing a voltage test at the module. Voltmeters that are used to perform this test should have high impedance in order to prevent damage to the module.

Answer D is incorrect. Both Technicians are correct.

TASK A.2.6

23. A vehicle has low power at all speeds. Technician A says that the engine exhaust system could be restricted. Technician B says that the torque converter stator could be defective. Who is correct?

A. A only

B. B only

C. Both A and B

D. Neither A nor B

Answer A is correct. Only Technician A is correct. A restricted exhaust system would cause the engine to have low power at all times. A backpressure test could be performed to test for this problem.

Answer B is incorrect. A torque converter could cause low power on takeoff, but would not cause low power at higher speeds.

Answer C is incorrect. Only Technician A is correct.

Answer D is incorrect. Technician A is correct.

C. Both A and B

D. Neither A nor B

Answer A is incorrect. Technician B is also correct.

Answer B is incorrect. Technician A is also correct.

Answer C is correct. Both Technicians are correct. An electronic transmission that stays in second gear all of the time likely is running in "limp mode." This is a protective mode that electronic transmissions have to prevent further damage when the computer senses a problem. A scan tool would be able to access the transmission computer trouble to retrieve possible trouble codes that could be causing the problem. In addition, the scan tool can override the transmission computer when the output test mode is chosen.

Answer D is incorrect. Both Technicians are correct.

25. A vehicle with an electronic transmission is being diagnosed for only having one forward gear and reverse gear. Technician A says that dirty transmission fluid could be the cause. Technician B says that a restricted cooler return pipe could be the cause. Who is correct?

TASK A.2.7

A. A only

B. B only

C. Both A and B

D. Neither A nor B

Answer A is incorrect. Dirty transmission fluid would not likely cause the vehicle to have no up-shift activity.

Answer B is incorrect. A restricted cooler return pipe would not cause the vehicle to have no up-shifts.

Answer C is incorrect. Neither Technician is correct.

Answer D is correct. Neither Technician is correct. Something is causing the vehicle to enter "limp mode." The transmission computer could be sensing a faulty sensor input or it may be detecting excess slippage in the gear train.

26. Technician A says that a misadjusted shift linkage could cause a no-crank condition. Technician B says that a range switch could cause a no-crank condition. Who is correct?

TASK B.1

A. A only

B. B only

C. Both A and B

D. Neither A nor B

Answer A is incorrect. Technician B is also correct.

Answer B is incorrect. Technician A is also correct.

Answer C is correct. Both Technicians are correct. A misadjusted shift linkage could cause the park/neutral switch or the range switch to not be in the correct position, which could cause a no-crank problem. The problem could also be caused by a faulty range switch.

Answer D is incorrect. Both Technicians are correct.

TASK B.2

27. Technician A says that the kick-down cable can be replaced without removing the transmission pan. Technician B says that the kick-down cable is not usually adjustable. Who is correct?

    A. A only

    B. B only

    C. Both A and B

    D. Neither A nor B

Answer A is correct. Only Technician A is correct. The transmission kick-down cable can usually be replaced without removing the transmission pan. The fastener that holds the cable housing to the transmission will have to be removed.

Answer B is incorrect. Most kick-down cables are adjustable.

Answer C is incorrect. Only Technician A is correct.

Answer D is incorrect. Technician A is correct.

TASK B.3

28. Transmission fluid is leaking from the input speed sensor area. Technician A says that the speed sensor o-ring can be replaced. Technician B says that the leak may be a cracked speed sensor. Who is correct?

    A. A only

    B. B only

    C. Both A and B

    D. Neither A nor B

Answer A is incorrect. Technician B is also correct.

Answer B is incorrect. Technician A is also correct.

Answer C is correct. Both Technicians are correct. A leak from the input speed area could be caused by either a faulty speed sensor o-ring or by a cracked speed sensor. The sensor could be removed to determine which problem exists.

Answer D is incorrect. Both Technicians are correct.

TASK B.5

29. Which of the following cooling system tests would be LEAST LIKELY to be performed on a late-model vehicle?

    A. Pressure test for leaks

    B. Freeze protection test with a refractometer

    C. Hose inspection

    D. Leak detection test with an electronic leak detector

Answer A is incorrect. It is a common practice to pressurize the cooling system on late-model vehicles to check for leaks. The system should not be pressurized above the radiator cap rating.

Answer B is incorrect. It is common to test the freeze protection of the coolant with a refractometer. This tool is the most accurate method testing the freeze protection of the coolant.

Answer C is incorrect. It is common to closely inspect the hoses on late-model vehicles. Performing this inspection on a routine basis will likely locate the source of leaks before they get dangerous.

Answer D is correct. It is not common to use any type of electronic leak-detecting device on the cooling system.

D.   Neither A nor B

Answer A is correct. Only Technician A is correct. Mineral spirits and compressed air would be a common method for cleaning the check balls from a valve body.

Answer B is incorrect. It is not a common practice to use sandpaper on the spool valves during the overhaul of a transmission. These devices are made with high levels of precision and should not be altered on a regular basis.

Answer C is incorrect. Only Technician A is correct.

Answer D is incorrect. Technician A is correct.

31.   All of the following statements about the engine cooling system are correct EXCEPT:

TASK B.5

A.   The boiling point is increased when the system builds pressure.
B.   The boiling point is decreased when the coolant is mixed with the water.
C.   The coolant recovery tank has a "full hot" and a "full cold" level.
D.   The freezing point is lowered when the coolant is mixed with the water.

Answer A is incorrect. The boiling point rises 3°F for each pound of pressure that is added to the cooling system.

Answer B is correct. The boiling point is increased when coolant is mixed with water.

Answer C is incorrect. There is a "full hot" and a "full cold" level on the recovery tank. The level of the coolant in this tank will vary as the engine temperature changes.

Answer D is incorrect. Mixing coolant with water lowers the freezing point to safe levels.

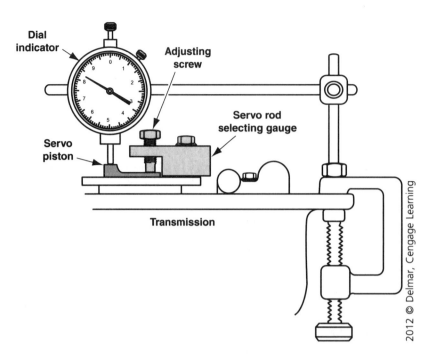

32. What process is taking place in the figure above?

TASK B.8

   A.   Testing the servo piston end-play

   B.   Testing the servo rod end-play

   C.   Selecting the correct servo apply pin

   D.   Selecting the correct servo piston

Answer A is incorrect. Technician B is also correct.

Answer B is incorrect. Technician A is also correct.

Answer C is correct. The process being displayed in the figure is the selection of the correct servo apply pin. Some transmissions do not have adjustable bands and the method for making sure that the band and servo will work correctly involves measuring the length of the servo apply pin. If the measurement is out of range, then a different length apply pin would be needed.

Answer D is incorrect.

33. What will be the most likely transmission condition if the servo piston fails to apply?

TASK B.8

   A.   The transmission will shift harshly in the gear that the servo is applied in.

   B.   The transmission will likely slip in the gear that the band is supposed to be applied in.

   C.   The vehicle will not move in forward or reverse gear.

   D.   The transmission will flare at the time that the band is commanded to be applied.

Answer A is incorrect. A failed accumulator could cause a harsh application of a band.

Answer B is correct. A vehicle with a servo piston problem would exhibit a problem of a transmission slip when the band is commanded to be applied.

Answer C is incorrect. An inoperative servo apply pin would not likely cause the transmission to not move forward or reverse.

Answer D is incorrect. The transmission would have more than a flare problem if the servo failed to apply.

Answer A is incorrect. Electrical tape is not a weather-tight method wire repair.

Answer B is incorrect. Scotch Lock connectors are not water resistant-style connectors.

Answer C is correct. Crimp-and-seal connectors used with a heat gun are very reliable and resist water intrusion well.

Answer D is incorrect. Butt connectors wrapped with electrical tape would not resist water intrusion.

35. Technician A says that a relay is used to control a high-level current circuit by using a small-current circuit. Technician B says that solenoids are used in electronic transmissions to control the transmission fluid pressure. Who is correct?

TASK B.9

 A. A only

 B. B only

 C. Both A and B

 D. Neither A nor B

Answer A is incorrect. Technician B is also correct.

Answer B is incorrect. Technician A is also correct.

Answer C is correct. Both Technicians are correct. Relays are electrical devices that are sometimes called magnetic switches. These devices have a coil that is controlled by a low-current circuit. When the coil is energized, the load contact closes and provides a high-current path to various load devices. Solenoids are very similar to relays but they have a moveable core. Solenoids are used in electronic transmissions to control fluid pressure.

Answer D is incorrect. Both Technicians are correct.

36. Which method would be most likely used when testing for live circuit voltage at a connector?

TASK B.9

 A. Piercing the wire insulation with a special tool

 B. Back-probing the connector with a T pin

 C. Disconnecting the connector to front-probe the connection

 D. Disconnecting the connector to back-probe the connection with a T pin

Answer A is incorrect. The wire insulation should not be pierced because water could enter at that point.

Answer B is correct. Back-probing a connector with a T pin will allow the technician to measure live voltage on the circuit without damaging the wires or connectors.

Answer C is incorrect. Disconnecting the connector causes a circuit interruption that is not recommended when troubleshooting active faults.

Answer D is incorrect. Disconnecting the connector causes a circuit interruption that is not recommended when troubleshooting active faults.

TASK B.11

37. Which method would be used to replace all of the transmission fluid?

    A. Transmission drain plug removal

    B. Bottom transmission pan removal

    C. Fluid exchange machine

    D. Cooler line removal

Answer A is incorrect. Most transmissions do not have a drain plug. Removing a plug would remove all of the fluid in the transmission, but not in the torque converter or transmission cooler and lines.

Answer B is incorrect. Removing the bottom transmission pan drains part of the transmission fluid, but a great deal of fluid will not be removed. The torque converter along with the transmission cooler and lines do not get drained when the bottom pan is removed.

Answer C is correct. A fluid exchange machine works well when the goal is to replace all of the transmission fluid.

Answer D is incorrect. Removing the cooler line is not recommended due to the potential of running the transmission fluid level too low.

TASK C.1.1

38. Technician A says that an engine support fixture should be installed on a vehicle prior to removing rear-wheel drive transmission. Technician B says that the drive shaft will have to be removed during the process of removing the transmission. Who is correct?

    A. A only

    B. B only

    C. Both A and B

    D. Neither A nor B

Answer A is incorrect. An engine support fixture is not usually needed when removing the transmission from a rear-wheel drive vehicle.

Answer B is correct. Only Technician B is correct. The drive shaft will have to be removed while removing a rear-wheel drive transmission.

Answer C is incorrect. Only Technician B is correct.

Answer D is incorrect. Technician B is correct.

TASK C.1.2

39. Technician A says that the torque converter can be easily forced into the transmission with the bell housing bolts while installing the transmission. Technician B says that the converter pilot should perfectly align with the crankshaft pilot bore during installation of the transmission. Who is correct?

    A. A only

    B. B only

    C. Both A and B

    D. Neither A nor B

Answer A is incorrect. The torque converter should never be forced into the transmission. The transmission can be internally damaged if this action is taken.

Answer B is correct. Only Technician B is correct. The converter pilot should align exactly with the crankshaft pilot bore during the installation of the transmission.

Answer C is incorrect. Only Technician B is correct.

Answer D is incorrect. Technician B is correct.

Answer A is incorrect. Adding clean transmission fluid to the torque converter is a common practice that is used when installing a torque converter.

Answer B is correct. The lockup clutch can't be tested before installing the torque converter.

Answer C is incorrect. The stator one-way clutch should be checked before installing the torque converter.

Answer D is incorrect. The torque converter end-play should be checked prior to installing the torque converter.

41. All of the following actions are required in order to check the flow rate of the transmission fluid cooler assembly EXCEPT:

    A. Drain the engine coolant.
    B. Remove the cooler return line.
    C. Start the engine and run at 1,000 rpm.
    D. Measure the fluid flow for 20 seconds.

TASK C.1.3

Answer A is correct. The engine coolant does not have to be drained in order to perform the cooler flow test.

Answer B is incorrect. The cooler return line needs to be removed in order to check the cooler flow rate.

Answer C is incorrect. The engine needs to be run at 1,000 rpm during the flow rate test.

Answer D is incorrect. The flow rate should be at monitored for 20 seconds. The cooler should flow at least one quart in 20 seconds.

42. All of the following statements are correct concerning bearing preload EXCEPT:

    A. Bearing preload is sometimes checked by testing the turning effort of a component.
    B. Bearing preload that is set too tight will loosen up once the unit heats up.
    C. Bearing preload that is set loose will tighten up as the unit heats up.
    D. Bearings must be lubricated in order to perform under an extended load.

TASK C.2.3

Answer A is incorrect. It is a common practice to check bearing preload by measuring the turning effort of the components involved.

Answer B is correct. Bearing preload that is set too tight will get tighter as the unit heats up due to the expansion characteristics of metal.

Answer C is incorrect. Metal tends to expand when the temperature rises, so bearing preload will usually get tighter when the temperature warms up.

Answer D is incorrect. Bearings need to be lubricated in order to perform under a load.

TASK C.2.6

43. An encapsulated check ball has a worn sleeve. Technician A says that a larger check ball should be installed. Technician B says that the check ball and sleeve can be removed and replaced. Who is correct?

   A.   A only
   B.   B only
   C.   Both A and B
   D.   Neither A nor B

   Answer A is incorrect. A larger check ball is not recommended when the sleeve gets worn. The assembly should be replaced.

   Answer B is correct. Only Technician B is correct. The check ball and sleeve should be removed and replaced when the sleeve gets worn.

   Answer C is incorrect. Only Technician B is correct.

   Answer D is incorrect. Technician B is correct.

TASK C.2.7

44. Technician A says that if a bushing shows wear it should be replaced. Technician B says that the shaft that mates with the bushing should also be checked if the bushing is worn. Who is correct?

   A.   A only
   B.   B only
   C.   Both A and B
   D.   Neither A nor B

   Answer A is incorrect. Technician B is also correct.

   Answer B is incorrect. Technician A is also correct.

   Answer C is correct. Both Technicians are correct. Any bushings that show wear in the transmission should be replaced. If a bushing is worn, the mating surface of that bushing could also show signs of wear and should be checked.

   Answer D is incorrect. Both Technicians are correct.

TASK C.2.9

45. A technician is diagnosing a transmission for a possible fractured case assembly. Technician A says that fluid can be poured into a suspect case bore to check for a leak. Technician B says that a fractured case can be repaired by welding the fractured area. Who is correct?

   A.   A only
   B.   B only
   C.   Both A and B
   D.   Neither A nor B

   Answer A is correct. Only Technician A is correct. Liquid is sometimes used to test for a fractured case. The passage should be blocked on one end and then the liquid can be poured into the case to see if it will leak off.

   Answer B is incorrect. It is not common to weld a fractured transmission case. The case should be replaced if it is fractured.

   Answer C is incorrect. Only Technician A is correct.

   Answer D is incorrect. Technician A is correct.

Answer A is incorrect. It is common to pour clean transmission fluid on the oil pump components during installation of the pump.

Answer B is correct. It is not a common practice to measure the pump internal gear with a micrometer. Feeler gauges are typically used to measure the clearance of the oil pump parts.

Answer C is incorrect. The oil pump bolts should be tightened to the correct torque using an accurate torque wrench.

Answer D is incorrect. It is a good practice to make sure the transmission pump will rotate after the fasteners have been tightened.

47. Technician A says that new friction discs should be soaked in engine oil prior to assembly of a clutch pack. Technician B says that the clutch pack clearance should be checked after building each clutch pack. Who is correct?

TASK C.3.1

A. A only
B. B only
C. Both A and B
D. Neither A nor B

Answer A is incorrect. Friction discs should be soaked in clean automatic transmission fluid prior to the assembly of a clutch pack.

Answer B is correct. Only Technician B is correct. It is vital to check the clutch pack clearance after the clutch pack has been assembled to make sure that the assembly is put together correctly.

Answer C is incorrect. Only Technician B is correct.

Answer D is incorrect. Technician B is correct.

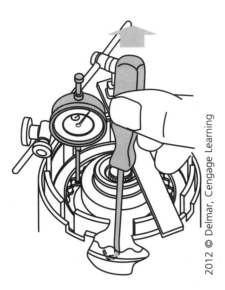

2012 © Delmar, Cengage Learning

TASK C.3.2

48. Referring to the figure above, Technician A says that this test will measure clutch pack clearance. Technician B says that air pressure should be applied to the clutch pack during this test. Who is correct?

A. A only

B. B only

C. Both A and B

D. Neither A nor B

Answer A is correct. Only Technician A is correct. The clutch pack clearance is being tested with a dial indicator. This method is a very accurate way of measuring clutch pack clearance.

Answer B is incorrect. The figure shows a pick being used to lift the apply plate. Air pressure is not needed when a pick is being used.

Answer C is incorrect. Only Technician A is correct.

Answer D is incorrect. Technician A is correct.

D.   Neither A nor B

Answer A is incorrect. Technician B is also correct.

Answer B is incorrect. Technician A is also correct.

Answer C is correct. Both Technicians are correct. Roller clutches should lock in one direction and freewheel in the other direction.

Answer D is incorrect. Both Technicians are correct.

50. What would be the most likely result of a transmission band that was adjusted too tightly?

TASK C.3.5

A.   Vehicle will take off in the wrong gear.
B.   Vehicle will not move forward or reverse.
C.   Vehicle will burn a clutch pack from overslipping.
D.   Servo seal will leak externally.

Answer A is correct. It is very likely that the vehicle will take off in the wrong gear in a vehicle that has a band that is adjusted too tightly.

Answer B is incorrect. A tight band would not likely cause the vehicle to lose all movement.

Answer C is incorrect. A tight band would not likely cause a burned clutch pack.

Answer D is incorrect. A tight band would not likely cause a servo seal to leak.

PREPARATION EXAM 6—ANSWER KEY

1.	C	**21.**	A	**41.**	A
2.	C	**22.**	C	**42.**	B
3.	C	**23.**	C	**43.**	C
4.	B	**24.**	B	**44.**	D
5.	B	**25.**	A	**45.**	A
6.	B	**26.**	B	**46.**	D
7.	A	**27.**	B	**47.**	B
8.	C	**28.**	A	**48.**	D
9.	B	**29.**	D	**49.**	C
10.	C	**30.**	A	**50.**	C
11.	B	**31.**	B		
12.	C	**32.**	C		
13.	D	**33.**	D		
14.	D	**34.**	A		
15.	B	**35.**	B		
16.	C	**36.**	D		
17.	C	**37.**	D		
18.	B	**38.**	C		
19.	C	**39.**	D		
20.	D	**40.**	A		

PREPARATION EXAM 6—EXPLANATIONS

TASK A.1.1

1. Which of the following conditions could cause the transmission to slip when the vehicle makes sharp turns?

   A. Brake switch is stuck open.

   B. Engine ignition system misfires.

   C. Transmission fluid is low.

   D. Transmission cooler is restricted.

   Answer A is incorrect. A stuck open brake switch could cause the stop lights to be inoperative. A stuck open brake switch could cause the cruise control and/or the torque converter clutch to be inoperative.

   Answer B is incorrect. An ignition system misfire could cause the vehicle to cut out under a load.

   Answer C is correct. Low transmission fluid would cause the transmission to slip when the vehicle makes sharp turns due to the fluid moving away from the pickup tube of the filter.

   Answer D is incorrect. A restricted transmission cooler would cause the transmission fluid to overheat, which could cause transmission component failure from high heat levels.

                                    Delmar, Cengage Learning ASE Test Preparation

D. Neither A nor B

Answer A is incorrect. Technician B is also correct.

Answer B is incorrect. Technician A is also correct.

Answer C is correct. Both Technicians are correct. Vacuum problems at the vacuum modulator could cause the transmission shifts to be firm while road testing. A plugged or leaking vacuum hose could cause the problem.

Answer D is incorrect. Both Technicians are correct.

3. A technician notices that the engine exhaust system is emitting white smoke during a road test. All of the following conditions could cause this condition EXCEPT:

TASK A.1.1

A. Engine intake gasket is blown.
B. Vacuum modulator is ruptured.
C. Transmission cooler is plugged.
D. Engine head gasket is blown.

Answer A is incorrect. A blown intake gasket could cause coolant to be ingested into the engine, which would cause white smoke to be emitted from the exhaust.

Answer B is incorrect. A ruptured vacuum modulator could cause transmission fluid to be pulled into the vacuum hose and be ingested into the engine, which would cause white smoke to be emitted from the exhaust.

Answer C is correct. A plugged transmission cooler would cause the transmission fluid to overheat and possibly cause component overheat.

Answer D is incorrect. A blown head gasket could cause coolant to be ingested into the engine, which would cause white smoke to be emitted from the exhaust.

4. Which of the following faults would be most likely to cause a clicking noise that increases when the transmission is shifted into drive and goes away when the transmission is shifted into park?

TASK A.1.2

A. Worn transmission pump
B. Cracked flexplate
C. Slipping transmission clutch pack
D. Torque converter lockup clutch slipping

Answer A is incorrect. A worn transmission pump would usually make a whining sound and it would not change when the vehicle is put into gear.

Answer B is correct. A cracked flexplate could cause a clicking sound that changes when the vehicle is shifted into drive due to the increased load.

Answer C is incorrect. A slipping clutch pack would cause the engine RPM to increase without an increase in torque output of the transmission.

Answer D is incorrect. A slipping torque converter lockup clutch would cause a shudder to occur at highway speeds.

5. A vehicle is being diagnosed for a power train vibration at 50 to 60 mph. Technician A says that a slipping clutch pack could cause this problem. Technician B says that a faulty universal joint could cause this problem. Who is correct?

    A. A only

    B. B only

    C. Both A and B

    D. Neither A nor B

    Answer A is incorrect. A slipping clutch pack could cause the engine RPM to increase without an increase in power output from the transmission.

    Answer B is correct. Only Technician B is correct. A faulty universal joint could cause the power train to have a vibration at highway speeds. The faulty joint would cause increased movement of the drive shaft, which would produce the vibration.

    Answer C is incorrect. Only Technician B is correct.

    Answer D is incorrect. Technician B is correct.

6. Which is the LEAST LIKELY factor that would be used to correctly check the transmission fluid?

    A. Vehicle on a level surface

    B. Engine at 1500 rpm

    C. Transmission in the correct gear position

    D. Transmission near operating temperature

    Answer A is incorrect. The vehicle needs to be on a level surface when correctly checking the transmission fluid.

    Answer B is correct. The engine speed does not need to be at 1500 rpm to correctly check the transmission fluid.

    Answer C is incorrect. The transmission needs to be in the correct gear when correctly checking the transmission fluid. Some transmissions are checked in park and some are checked in neutral.

    Answer D is incorrect. The transmission needs to be near normal operating temperature to correctly check the transmission fluid.

7. A vehicle is being diagnosed that has extremely burned and discolored transmission fluid. Technician A says that burned clutch discs in a clutch pack could be the cause. Technician B says that a leak in the transmission cooler could be the cause. Who is correct?

    A. A only

    B. B only

    C. Both A and B

    D. Neither A nor B

    Answer A is correct. Only Technician A is correct. A clutch pack that has burned clutch discs would cause the transmission fluid to be very dark and discolored.

    Answer B is incorrect. A leak in the transmission cooler could cause engine coolant to enter the transmission, which would cause the fluid to have a brown or pink color. The cooler could also allow transmission fluid to be pumped into the engine cooling system.

    Answer C is incorrect. Only Technician A is correct.

    Answer D is incorrect. Technician A is correct.

C. Both A and B

D. Neither A nor B

Answer A is incorrect. Technician B is also correct.

Answer B is incorrect. Technician A is also correct.

Answer C is correct. Both Technicians are correct. A partially block governor passage could cause late up-shifts. In addition, a sticking second gear shift valve could cause a delayed shift into the second gear.

Answer D is incorrect. Both Technicians are correct.

9. Technician A says that a low stall speed could be caused by a slipping clutch pack assembly. Technician B says that a high stall speed could be caused by a slipping band. Who is correct?

TASK A.1.5

A. A only

B. B only

C. Both A and B

D. Neither A nor B

Answer A is incorrect. A slipping clutch pack would cause a high stall speed.

Answer B is correct. Only Technician B is correct. A high stall speed could be caused by a slipping band if the band was commanded to be on during the stall test.

Answer C is incorrect. Only Technician B is correct.

Answer D is incorrect. Technician B is correct.

10. Technician A says that the transmission computer controls the torque converter lockup clutch. Technician B says that an inoperative torque converter lockup clutch can cause a decrease in fuel economy. Who is correct?

TASK A.1.6

A. A only

B. B only

C. Both A and B

D. Neither A nor B

Answer A is incorrect. Technician B is also correct.

Answer B is incorrect. Technician A is also correct.

Answer C is correct. Both Technicians are correct. The transmission computer or the power train control module controls the operation of the torque converter clutch. Since the torque converter lockup clutch reduces the engine RPM when it engages, fuel economy would be reduced if the converter clutch was inoperative.

Answer D is incorrect. Both Technicians are correct.

11. An inoperative torque converter lockup clutch will cause all of the following results EXCEPT:

A. Reduced fuel economy

B. Increased engine temperature

C. Increased engine RPM at highway speeds

D. Diagnostic trouble code

Answer A is incorrect. The fuel economy will be reduced on a vehicle with an inoperative torque converter lockup clutch because the engine will turn at more RPM at highway speeds.

Answer B is correct. An inoperative torque converter lockup clutch would not have an effect on the engine temperature.

Answer C is incorrect. An inoperative torque converter lockup clutch would cause increased engine RPM at highway speeds.

Answer D is incorrect. An inoperative torque converter lockup clutch would likely cause a diagnostic trouble code (DTC) in the transmission computer or the power train control module.

12. Which of the following items should be monitored during the road test of a vehicle with an electronic transmission?

A. Engine oil temperature

B. Speed sensor resistance

C. Transmission up-shift points

D. Transmission fluid level

Answer A is incorrect. It is not normal to monitor engine oil temperature during the road test for a transmission concern.

Answer B is incorrect. It would not be possible to monitor the resistance of the speed sensor during the road test.

Answer C is correct. It is a good practice to monitor the up-shift points of a transmission during the road test. The up-shift points will vary depending how aggressively the vehicle is being driven.

Answer D is incorrect. It would not be possible to monitor the fluid level during the road test. The fluid level could be checked prior to beginning the road test to make sure there was adequate fluid to safely perform the test.

13. Technician A says that the governor pressure should be monitored during the road test of a vehicle with an electronic transmission. Technician B says that the vacuum should be checked at the modulator valve during the road test of a vehicle with an electronic transmission. Who is correct?

A. A only

B. B only

C. Both A and B

D. Neither A nor B

Answer A is incorrect. Electronic transmissions do not use a governor.

Answer B is incorrect. Electronic transmissions do not typically use a vacuum modulator.

Answer C is incorrect. Neither Technician is correct.

Answer D is correct. Neither Technician is correct. Electronic transmissions use sensors to perform the jobs that the governor and modulator performed on hydraulically shifted transmissions. Speed sensors are used as a vehicle speed input and the MAP (vacuum) sensor is used as input for engine load.

Answer A is incorrect. Excessive line pressure would cause very harsh shifts.

Answer B is incorrect. Low line pressure would cause soft shifts and possibly even a shift that slips.

Answer C is incorrect. Neither Technician is correct.

Answer D is correct. Neither Technician is correct. Excessive line pressure would cause the shifts to be very firm and low line pressure would cause the shifts to be very soft.

15. A vehicle with an electronic transmission has been tested for transmission pump output pressure. The pressures attained during the test were above the specifications in all ranges. Which of the following conditions would be most likely the cause of this problem?

TASK A.2.2

   A.   Restricted transmission filter

   B.   Loose terminal connection at the pressure control solenoid

   C.   Punctured transmission fluid pickup tube

   D.   Excess transmission pump clearance

Answer A is incorrect. A restricted transmission filter would cause lower line pressure.

Answer B is correct. A loose terminal at the pressure control solenoid would cause the pressure to be higher than normal due to the decreased current flow to the solenoid.

Answer C is incorrect. A punctured transmission fluid pickup tube would cause reduced transmission line pressure.

Answer D is incorrect. Excess transmission pump clearance would cause lower line pressure.

16. A vehicle is being diagnosed for a problem of the engine dying when coming to a stop. After restarting the engine, it again dies when the transmission is shifted into any gear. Technician A says that the brake switch could be the cause. Technician B says that the torque converter clutch solenoid could be the cause. Who is correct?

TASK A.2.3

   A.   A only

   B.   B only

   C.   Both A and B

   D.   Neither A nor B

Answer A is incorrect. A failed brake switch could cause the vehicle to die when coming to a stop, but the engine would not die after being restarted and put back into gear.

Answer B is correct. Only Technician B is correct. A torque converter clutch solenoid could cause the torque converter clutch to stick in the applied position which would cause the engine to die when the vehicle is stopped while in gear.

Answer C is incorrect. Only Technician B is correct.

Answer D is incorrect. Technician B is correct.

TASK A.2.3

17. All of the following conditions have to be met in order for the torque converter clutch to engage EXCEPT:

A. Light to moderate throttle application

B. Vehicle above the minimum converter clutch set speed

C. Transmission in second gear

D. Engine temperature minimum met

Answer A is incorrect. The throttle application should be light to moderate in order for the torque converter clutch to engage.

Answer B is incorrect. The vehicle must be traveling above the minimum converter clutch set speed.

Answer C is correct. The torque converter will not typically be engaged in second gear. The torque converter clutch is usually engaged in the top two gears.

Answer D is incorrect. The engine temperature should be at least at the minimum level in order for the torque converter clutch to engage.

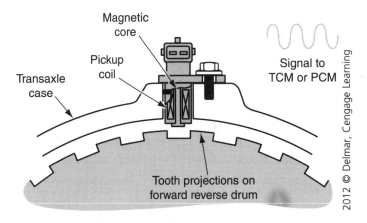

18. Referring to the figure above, all of the following statements are correct about the speed sensor above EXCEPT:

TASK A.2.4

A. The speed sensor contains a coil.

B. The output frequency of the sensor increases as the forward/reverse drum slows down.

C. The output voltage of the sensor increases as the forward/reverse drum speeds up.

D. The speed sensor contains a permanent magnet.

Answer A is incorrect. The speed sensor is a permanent magnet device that contains a coil that is wrapped around a permanent magnet.

Answer B is correct. The frequency of the sensor should decrease as the forward/reverse drum slows down.

Answer C is incorrect. The output voltage should increase as the speed of the forward/reverse drum increases.

Answer D is incorrect. The speed sensor is a permanent magnet device that contains a coil that is wrapped around a permanent magnet.

Answer A is incorrect. A database or a service manual would be needed to find a flowchart for a trouble code.

Answer B is incorrect. A database or a service manual would be needed for following the fluid flow in a hydraulic circuit.

Answer C is correct. A wire schematic could be used for determining the pin identification for a connector on an electronic transmission. Wire schematics also show wire colors, circuit numbers, power distribution, and ground distribution for electronic circuits.

Answer D is incorrect. The wiring diagram does not reveal the physical location of the transmission electronic components. A component locator is used to find this information.

20. Technician A says that a 12 volt battery that has 6 volts at the posts is 50 percent charged. Technician B says that a 12 volt battery that has 12.6 volts at the posts is overcharged. Who is correct?

TASK A.2.5

    A.   A only

    B.   B only

    C.   Both A and B

    D.   Neither A nor B

Answer A is incorrect. The 50 percent charge level on a 12 volt battery is 12.2 volts.

Answer B is incorrect. A battery with 12.6 volts at the posts is fully charged.

Answer C is incorrect. Neither Technician is correct.

Answer D is correct. Neither Technician is correct. A battery that has only 6 volts at the posts is severely discharged. A fully charged automotive battery should have 12.6 volts. These batteries have 6 cells that produce 2.1 volts each.

21. A vehicle is being diagnosed with a transmission that stays in second gear all of the time. The supply (B+) voltage was tested at the transmission computer and was found to be at 10.5 volts with the engine running. Technician A says that a voltage drop test should be performed on the (B+) circuit. Technician B says that the problem could be caused by a wire repair made with wire that is too large. Who is correct?

TASK A.2.5

    A.   A only

    B.   B only

    C.   Both A and B

    D.   Neither A nor B

Answer A is correct. Only Technician A is correct. It would be advisable to perform a voltage drop test when the supply voltage is lower than specifications.

Answer B is incorrect. A wire repair using a larger wire would never cause a reduction of supply voltage.

Answer C is incorrect. Only Technician A is correct.

Answer D is incorrect. Technician A is correct.

TASK A.2.6

22. The transmission fluid is found to be overfilled and is a pink color. Which of the following conditions could cause this problem?

   A. Engine head gasket is leaking.

   B. Engine intake gasket is leaking.

   C. Transmission cooler is leaking.

   D. Transmission solenoid pack gasket is leaking.

   Answer A is incorrect. A leaking engine head gasket could cause coolant to leak externally or be pulled into the combustion chamber of the engine.

   Answer B is incorrect. A leaking intake manifold gasket could cause the engine to pull coolant into the combustion chamber and then make the exhaust white in color.

   Answer C is correct. A leaking transmission cooler would cause the transmission fluid to turn pink due to the engine coolant being mixed with the transmission fluid. This would also cause the fluid level to become overfilled.

   Answer D is incorrect. A leaking solenoid pack would cause transmission fluid to leak externally and eventually cause a low fluid level in the transmission.

TASK A.2.6

23. The coolant reserve container is overfilled and seems to be red in color. Which of the following conditions could cause this problem?

   A. Transmission solenoid pack gasket is leaking.

   B. Engine intake gasket is leaking.

   C. Transmission cooler is leaking.

   D. Transmission modulator valve is leaking.

   Answer A is incorrect. A leaking solenoid pack would cause an external transmission fluid leak that would be noticeable on the transmission case.

   Answer B is incorrect. A leaking engine intake gasket could cause coolant to leak externally or it could be pulled into the engine combustion chamber.

   Answer C is correct. A leaking transmission cooler could cause transmission fluid to be pumped into the cooling system.

   Answer D is incorrect. A leaking modulator valve would cause an external transmission fluid leak or possibly a problem of the engine producing white smoke, caused by the engine pulling transmission fluid through the vacuum hose.

Answer A is incorrect. A broken input shaft could cause the transmission to be completely motionless due to the main input into the transmission being inoperative.

Answer B is correct. A faulty input shaft speed sensor could cause the transmission to go into "limp mode," but it would still have one forward gear along with reverse.

Answer C is incorrect. Transmission fluid level that is so low that the filter will not pick it up would cause the transmission to not have forward or reverse.

Answer D is incorrect. A faulty transmission oil pump could cause the transmission to not have forward or reverse capability.

25. The main power fuse for the electronic control system has failed. Technician A says that the vehicle will still have one forward gear as well as reverse. Technician B says that this fault should set a diagnostic trouble code (DTC) in the transmission computer. Who is correct?

TASK A.2.7

    A.   A only

    B.   B only

    C.   Both A and B

    D.   Neither A nor B

Answer A is correct. Only Technician A is correct. An electronic transmission will still have one forward gear along with reverse if all of the electrical power is lost to the transmission.

Answer B is incorrect. A failed main power fuse for the electronic control system would not likely set a diagnostic trouble code due to the lack of voltage supplied to the computer.

Answer C is incorrect. Only Technician A is correct.

Answer D is incorrect. Technician A is correct.

26. Which of the following components could cause the electronic shift indicator to show the incorrect gear?

TASK B.1

    A.   Faulty park/neutral switch

    B.   Misadjusted range sensor

    C.   Misadjusted throttle position sensor

    D.   Faulty speed sensor

Answer A is incorrect. A faulty park/neutral switch could cause a no-crank condition.

Answer B is correct. A misadjusted range sensor could cause the shift indicator to show the incorrect gear because this sensor is used as an input about the gear range to the transmission computer.

Answer C is incorrect. A misadjusted throttle position sensor could cause the shift points to be altered as well as engine performance problems.

Answer D is incorrect. A faulty speed sensor could cause the transmission to go into "limp mode."

TASK B.1

27. Technician A says that the bottom pan must be removed in order to replace the park/neutral switch. Technician B says that the range sensor can be replaced without removing the transmission pan. Who is correct?

    A.   A only

    B.   B only

    C.   Both A and B

    D.   Neither A nor B

Answer A is incorrect. The park/neutral switch can usually be replaced without removing the transmission pan.

Answer B is correct. Only Technician B is correct. The range sensor can usually be replaced without removing the transmission pan. This sensor is typically mounted near the transmission shift linkage.

Answer C is incorrect. Only Technician B is correct.

Answer D is incorrect. Technician B is correct.

TASK B.2

28. The throttle cable needs to be replaced on a late-model vehicle. Technician A says that the kick-down cable may need to be adjusted after installing the new throttle cable. Technician B says that the throttle plate is typically spring loaded to the open position. Who is correct?

    A.   A only

    B.   B only

    C.   Both A and B

    D.   Neither A nor B

Answer A is correct. Only Technician A is correct. It may be necessary to adjust the kick-down cable after a new throttle cable is installed because the new cable is slightly different from the old cable.

Answer B is incorrect. The throttle plate is spring loaded to the closed position.

Answer C is incorrect. Only Technician A is correct.

Answer D is incorrect. Technician A is correct.

TASK B.2

29. Which of the following statements best describes the action of the throttle valve linkage?

    A.   The throttle valve linkage connects the engine to the cruise control servo.

    B.   The throttle valve linkage connects the throttle pedal to the engine.

    C.   The throttle valve provides vehicle speed feedback to the transmission valve body.

    D.   The throttle valve provides engine load feedback to the transmission valve body.

Answer A is incorrect. The cruise control cable connects the engine to the cruise control servo.

Answer B is incorrect. The accelerator cable/linkage connects the throttle pedal to the engine throttle plate.

Answer C is incorrect. The governor valve provides vehicle speed feedback to the transmission valve body.

Answer D is correct. The throttle valve linkage provides engine load feedback to the transmission valve body. When the throttle is applied heavily, the valve body will delay the shifts until a higher speed. When the throttle is applied lightly, the valve body will allow the shift points to occur at a lower speed.

D.   Neither A nor B

Answer A is correct. Only Technician A is correct. A leak in the output speed sensor area could be caused by a damaged speed sensor o-ring.

Answer B is incorrect. It may not be necessary to replace the speed sensor for a leak. The o-ring could be the cause and need to be replaced.

Answer C is incorrect. Only Technician A is correct.

Answer D is incorrect. Technician A is correct.

31. Transmission fluid is leaking from the manual linkage near the transmission case. Technician A says that the transmission pan will need to be removed to replace the shift linkage seal. Technician B says that the shift linkage will need to be disassembled to replace the shift linkage seal. Who is correct?

TASK B.3

A.   A only

B.   B only

C.   Both A and B

D.   Neither A nor B

Answer A is incorrect. The transmission pan does not have to be removed in order to replace the shift linkage metal-clad seal. This repair can be performed on some transmissions by removing the linkage from the outside of the transmission.

Answer B is correct. Only Technician B is correct. The shift linkage seal can be replaced by disassembling the shift linkage.

Answer C is incorrect. Only Technician B is correct.

Answer D is incorrect. Technician B is correct.

32. A late-model vehicle is being diagnosed for a cooling system leak. Technician A says that applying pressure to the system with a pressure tester may be necessary. Technician B says that adding coolant dye to the system may be necessary. Who is correct?

TASK B.5

A.   A only

B.   B only

C.   Both A and B

D.   Neither A nor B

Answer A is incorrect. Technician B is also correct.

Answer B is incorrect. Technician A is also correct.

Answer C is correct. Both Technicians are correct. Applying pressure with a pressure tester may be necessary to find some cooling system leaks. In addition, the technician may need to add dye to the coolant in order to locate the source of challenging cooling system leaks.

Answer D is incorrect. Both Technicians are correct.

TASK B.6

33. Technician A says that the valve body mating surface should be cleaned with medium grit sandpaper. Technician B says that the valve body bores should be cleaned with a light shop rag. Who is correct?

    A. A only

    B. B only

    C. Both A and B

    D. Neither A nor B

    Answer A is incorrect. Sandpaper should never be used on a valve body mating surface.

    Answer B is incorrect. Shop rags should never be used on the valve body because the possibility of a piece of lint getting trapped in the bore.

    Answer C is incorrect. Neither Technician is correct.

    Answer D is correct. Neither Technician is correct. The valve body mating surface should never be sanded or damaged with any surface prep tools. Shop rags should never be used when cleaning the valve body. Compressed air can be used to clean the valve body without causing damage.

TASK B.7

34. Which of the following tools would be LEAST LIKELY to be acceptable when tightening the bolts of a valve body?

    A. 3/4 drive foot/pound torque wrench

    B. Inch/pound torque wrench

    C. Newton/meter torque wrench

    D. Small range foot/pound torque wrench

    Answer A is correct. A 3/4 drive foot/pound torque wrench would not commonly be used to tighten the bolts of a valve body. This type of tool would not be accurate enough for the small and precise torque range that valve bodies require.

    Answer B is incorrect. Inch/pound torque wrenches are often used to tighten the bolts of a valve body.

    Answer C is incorrect. A newton/meter torque wrench is sometimes used to tighten the bolts of a valve body.

    Answer D is incorrect. A small range foot/pound torque wrench is sometimes used to tighten the bolts of a valve body.

TASK B.8

35. A second-gear servo is jammed in the applied position. What is the most likely transmission result?

    A. The vehicle will never shift to second gear.

    B. The vehicle will launch from a stop in second gear.

    C. The vehicle will be stuck in reverse gear.

    D. The vehicle will not pull in forward or reverse.

    Answer A is incorrect. The vehicle would take off in second gear if the servo is jammed in the applied position.

    Answer B is correct. The vehicle will likely take off on second gear if the second gear servo is jammed in the applied position.

    Answer C is incorrect. The vehicle would likely bind up when the shifter is moved to reverse.

    Answer D is incorrect. The vehicle would likely launch from a stop in second gear and then bind up when the shifter is moved to reverse.

Answer A is incorrect. A chipped planetary gear would likely cause a whining sound at various vehicle speeds.

Answer B is incorrect. A restricted transmission cooler would cause the transmission fluid to overheat under high load conditions. This problem could potentially cause component overheating in the transmission drive train.

Answer C is incorrect. A loosely adjusted band would cause the transmission to slip when the band was applied.

Answer D is correct. A broken rear transmission mount could cause a bumping noise when backing up a steep incline due to the excess movement allowed by the broken mount.

37. Where would the transmission filter be LEAST LIKELY located?

TASK B.11

   A. Mounted in a transmission cooler line

   B. Mounted on the transmission case

   C. Mounted inside the transmission pan

   D. Mounted inside the torque converter

Answer A is incorrect. Some transmissions have inline transmission filters mounted in the cooler lines.

Answer B is incorrect. Some transmissions have a removable transmission filter mounted on the transmission case.

Answer C is incorrect. Many transmissions mount the transmission filter inside the transmission pan.

Answer D is correct. The transmission filter is not mounted inside the torque converter.

38. Technician says the front pump seal should be replaced any time that the torque converter is replaced. Technician B says that a new torque converter should be at least partly filled with new fluid prior to installing into the transmission. Who is correct?

TASK C.1.1

   A. A only

   B. B only

   C. Both A and B

   D. Neither A nor B

Answer A is incorrect. Technician B is also correct.

Answer B is incorrect. Technician A is also correct.

Answer C is correct. Both Technicians are correct. It is a good practice to replace the front pump seal any time that the torque converter is replaced. It is also a good practice to add new transmission fluid to the new torque converter prior to installing into the transmission.

Answer D is incorrect. Both Technicians are correct.

TASK C.1.2

39. Failing to completely install the torque converter into the transmission prior to installing the transmission will most likely cause which of the following results?

   A. Broken input shaft
   B. Front crankshaft seal to leak
   C. Damage to the clutch pack piston
   D. Damage to the transmission oil pump gears

   Answer A is incorrect. Installing an imbalanced torque converter could cause the input shaft to break.

   Answer B is incorrect. The front crankshaft seal may leak if the engine PCV valve is plugged.

   Answer C is incorrect. Installing the piston return spring incorrectly could cause damage to the clutch pack piston.

   Answer D is correct. The transmission oil pump gears will likely get damaged if the torque converter does not get properly installed into the transmission prior to installing the transmission into the vehicle.

TASK C.1.3

40. Technician A says that the transmission cooler should be flushed every time that the transmission is rebuilt or replaced. Technician B says that the cooler should be flushed every 30,000 miles. Who is correct?

   A. A only
   B. B only
   C. Both A and B
   D. Neither A nor B

   Answer A is correct. Only Technician A is correct. It is a good practice to flush the transmission cooler each time that the transmission is rebuilt or replaced.

   Answer B is incorrect. It is not a normal practice to flush the transmission cooler every 30,000 miles.

   Answer C is incorrect. Only Technician A is correct.

   Answer D is incorrect. Technician A is correct.

TASK C.1.3

41. Which of the following actions would be LEAST LIKELY to be performed by the technician while running a transmission fluid cooler flow test?

   A. Remove the cooler supply line.
   B. Remove the cooler return line.
   C. Start the engine and run at 1,000 rpm.
   D. Measure the fluid flow for 20 seconds.

   Answer A is correct. The cooler supply line should not be removed while performing a cooler flow test.

   Answer B is incorrect. The cooler return line should be removed while performing a cooler flow test.

   Answer C is incorrect. The engine should be run at 1,000 rpm while performing a cooler flow test.

   Answer D is incorrect. The cooler flow test is typically run for 20 seconds. One quart of fluid should be pumped out into the catch container in 20 seconds.

D. Neither A nor B

Answer A is incorrect. It is not a common practice to replace all bushings during each transmission overhaul operation.

Answer B is correct. Only Technician B is correct. The material that the bushings are made from is typically softer than the shaft that rides against the bushing.

Answer C is incorrect. Only Technician B is correct.

Answer D is incorrect. Technician B is correct.

43. The planetary gear set is blue and discolored. Technician A says that the gear set has been overheated. Technician B says that the gear set should be replaced. Who is correct?

TASK C.2.8

    A.   A only

    B.   B only

    C.   Both A and B

    D.   Neither A nor B

Answer A is incorrect. Technician B is also correct.

Answer B is incorrect. Technician A is also correct.

Answer C is correct. Both Technicians are correct. Blue and discolored components indicate that the component has been hot and overheated. Overheated components should be replaced, as the metal could be weakened by overheating.

Answer D is incorrect. Both Technicians are correct.

TASK C.2.10

44. Technician A says that a transaxle drive chain should be replaced during every overhaul procedure. Technician B says that the chain should be greased heavily during assembly to prevent early chain failure. Who is correct?

    A. A only

    B. B only

    C. Both A and B

    D. Neither A nor B

    Answer A is incorrect. It is not a common practice to replace the transaxle drive chain during every overhaul procedure.

    Answer B is incorrect. It is not a common practice to apply grease to the drive chain because it receives lubrication from the transmission fluid.

    Answer C is incorrect. Neither Technician is correct.

    Answer D is correct. Neither Technician is correct.

TASK C.2.11

45. Which of the following methods would most likely be used when testing the contact pattern on the final drive gears in a transaxle?

    A. Gear compound

    B. Slide caliper

    C. Outside micrometer

    D. Feeler gauge

    Answer A is correct. Gear compound is often used when testing the contact pattern of the final drive gears in a transaxle. This substance provides good feedback about where the teeth are touching each other.

    Answer B is incorrect. A slide caliper is used to make various transmission measurements such as outside diameter and inside diameter, as well as depth.

    Answer C is incorrect. Outside micrometers are used to measure the outside diameter of components in the transmission.

    Answer D is incorrect. A feeler gauge is used to check clutch pack clearance.

Answer A is incorrect. Clutches and bands should be installed in the transmission after they have been soaked for at least 30 minutes in ATF.

Answer B is incorrect. Clutches and bands should be installed in the transmission after they have been soaked for at least 30 minutes in ATF.

Answer C is incorrect. Neither Technician is correct.

Answer D is correct. Neither Technician is correct. Clutches and bands should be installed in the transmission after they have been soaked for at least 30 minutes in ATF.

47. All of the following actions should be made prior to the assembly of a clutch pack EXCEPT:

    A. The friction discs should be soaked in clean transmission fluid.
    B. The clutch pack apply plate should be machined on a lathe to polish the surface.
    C. The clutch pack drum should be thoroughly cleaned and inspected.
    D. The clutch pack piston seals should be installed and lubricated with assembly lube.

TASK C.3.1

Answer A is incorrect. It is a common practice to soak the friction discs in clean transmission fluid prior to the assembly of a clutch pack.

Answer B is correct. The apply plate is not typically machined on a lathe prior to assembling a clutch pack.

Answer C is incorrect. The clutch pack drum should always be thoroughly cleaned and inspected prior to assembling the clutch pack.

Answer D is incorrect. The clutch pack piston seals should be installed and lubricated with assembly lube prior to assembling the clutch pack.

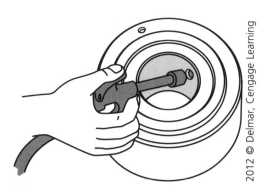

2012 © Delmar, Cengage Learning

TASK C.3.3

48. The test being performed in the figure above will provide feedback to the technician about all of the following clutch pack components EXCEPT:

A. Clutch pack piston

B. Clutch pack piston seal

C. Piston return spring

D. Clutch pack apply plate

Answer A is incorrect. The clutch pack air test will test the clutch pack piston.

Answer B is incorrect. The clutch pack air test will test the clutch pack piston seal.

Answer C is incorrect. The clutch pack air test will test the piston return spring.

Answer D is correct. The clutch pack air test will not provide any feedback about the clutch pack apply plate. The plate clearance would need to be tested and then the apply plate would need to be measured with a caliper or a micrometer to select the correct apply plate.

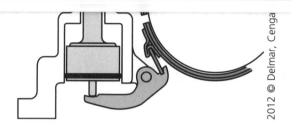

2012 © Delmar, Cenga

49. All of the following components are present in the figure above EXCEPT:

    A. Transmission band adjusting bolt and nut
    B. Transmission band
    C. Valve body
    D. Transmission servo

TASK C.3.5

Answer A is incorrect. The figure of the apply servo and band components includes the band adjusting bolt and nut.

Answer B is incorrect. The figure includes the transmission band.

Answer C is correct. The figure of the apply servo and band components does not include the valve body.

Answer D is incorrect. The figure includes the apply servo.

50. What would be the most likely result of a transmission band that was adjusted too loosely?

    A. Vehicle will take off in the wrong gear.
    B. Vehicle will not move forward or reverse.
    C. The transmission will have a slip or flare when the band is applied.
    D. The servo seal will leak externally.

TASK C.3.5

Answer A is incorrect. A band that was adjusted too tightly would cause the transmission to take off in the wrong gear.

Answer B is incorrect. A broken input shaft or a failed transmission pump could cause the transmission to lack forward or reverse gears.

Answer C is correct. A loosely adjusted band would cause the transmission to slip or flare when the band is applied.

Answer D is incorrect. A cut servo seal would cause the servo to leak externally.

PREPARATION EXAM ANSWER SHEET FORMS

ANSWER SHEET

1. _____	21. _____	41. _____
2. _____	22. _____	42. _____
3. _____	23. _____	43. _____
4. _____	24. _____	44. _____
5. _____	25. _____	45. _____
6. _____	26. _____	46. _____
7. _____	27. _____	47. _____
8. _____	28. _____	48. _____
9. _____	29. _____	49. _____
10. _____	30. _____	50. _____
11. _____	31. _____	
12. _____	32. _____	
13. _____	33. _____	
14. _____	34. _____	
15. _____	35. _____	
16. _____	36. _____	
17. _____	37. _____	
18. _____	38. _____	
19. _____	39. _____	
20. _____	40. _____	

4. _____	24. _____	44. _____
5. _____	25. _____	45. _____
6. _____	26. _____	46. _____
7. _____	27. _____	47. _____
8. _____	28. _____	48. _____
9. _____	29. _____	49. _____
10. _____	30. _____	50. _____
11. _____	31. _____	
12. _____	32. _____	
13. _____	33. _____	
14. _____	34. _____	
15. _____	35. _____	
16. _____	36. _____	
17. _____	37. _____	
18. _____	38. _____	
19. _____	39. _____	
20. _____	40. _____	

ANSWER SHEET

1. _____	21. _____	41. _____
2. _____	22. _____	42. _____
3. _____	23. _____	43. _____
4. _____	24. _____	44. _____
5. _____	25. _____	45. _____
6. _____	26. _____	46. _____
7. _____	27. _____	47. _____
8. _____	28. _____	48. _____
9. _____	29. _____	49. _____
10. _____	30. _____	50. _____
11. _____	31. _____	
12. _____	32. _____	
13. _____	33. _____	
14. _____	34. _____	
15. _____	35. _____	
16. _____	36. _____	
17. _____	37. _____	
18. _____	38. _____	
19. _____	39. _____	
20. _____	40. _____	

4. _____
5. _____
6. _____
7. _____
8. _____
9. _____
10. _____
11. _____
12. _____
13. _____
14. _____
15. _____
16. _____
17. _____
18. _____
19. _____
20. _____

24. _____
25. _____
26. _____
27. _____
28. _____
29. _____
30. _____
31. _____
32. _____
33. _____
34. _____
35. _____
36. _____
37. _____
38. _____
39. _____
40. _____

44. _____
45. _____
46. _____
47. _____
48. _____
49. _____
50. _____

ANSWER SHEET

1. _____	21. _____	41. _____
2. _____	22. _____	42. _____
3. _____	23. _____	43. _____
4. _____	24. _____	44. _____
5. _____	25. _____	45. _____
6. _____	26. _____	46. _____
7. _____	27. _____	47. _____
8. _____	28. _____	48. _____
9. _____	29. _____	49. _____
10. _____	30. _____	50. _____
11. _____	31. _____	
12. _____	32. _____	
13. _____	33. _____	
14. _____	34. _____	
15. _____	35. _____	
16. _____	36. _____	
17. _____	37. _____	
18. _____	38. _____	
19. _____	39. _____	
20. _____	40. _____	

4. _____
5. _____
6. _____
7. _____
8. _____
9. _____
10. _____
11. _____
12. _____
13. _____
14. _____
15. _____
16. _____
17. _____
18. _____
19. _____
20. _____

24. _____
25. _____
26. _____
27. _____
28. _____
29. _____
30. _____
31. _____
32. _____
33. _____
34. _____
35. _____
36. _____
37. _____
38. _____
39. _____
40. _____

44. _____
45. _____
46. _____
47. _____
48. _____
49. _____
50. _____

Glossary

AC An acronym for (1) alternating current or (2) air conditioning. A/C is also a commonly used acronym for air conditioning.

Aerated The introduction of air bubbles into the transmission fluid causing the fluid to expand and become compressible.

Alternating Current (AC) The type of electrical current produced in an alternator.

Anodized To coat with electrolysis a metal surface with a protective oxide.

Arcing A term that applies to the spark that occurs when electricity jumps a gap.

Battery A device that stores electrical energy in chemical form.

Bench Test A test or series of tests done by the technician on the work bench rather than on the vehicle.

Brakes The system that slows or stops a vehicle.

Bushing A sleeve, usually bronze, inserted into a bore to serve as a bearing for a rotating member.

Case Grounded A term used when a component is grounded electrically through its case and is fastened directly to a metal, grounded part of the vehicle.

Chocked A term used when the wheels are blocked with safety chocks to insure safe working conditions on a running vehicle.

Clutch Pack A series of clutch discs and plates that act as a driving or driven unit.

Computer An electronic device capable of following instructions and performing operations without human intervention.

Constant-Velocity Joint (CV Joint) One or more universal joints specially designed so acceleration-deceleration effects cancel each other out.

Control Circuit A circuit that has the actual on/off responsibility in a circuit. It could be on the positive side or the negative side of the load, switch, or relay.

Coolant A fluid, usually a mixture of water and antifreeze, used in the cooling system.

Crankshaft A revolving part in the lower section of the engine to which the connecting rods are attached.

DC An acronym for direct current.

Direct Current A form of electrical energy produced by a battery.

Dressed A term used for the process of slightly altering a surface. Dressing can make a surface slightly smoother, rougher, or it can just remove slight imperfections.

Electrical A type of power produced by alternating or direct current.

Electrical Harness A group of wires usually terminating at one or more connectors. Used to connect various components and systems.

Electronic That branch of science dealing with the motion, behavior, and emission of currents of free electrons.

Electronic-Control Computer A digital device that controls engine and transmission functions electrically.

Electronic Controls A term used for the electronic modules and associated components used to control many of the systems on today's vehicles.

Electronic Transmission An automatic transmission that is computer controlled.

Elongated The stretching of the parts so they are no longer as designed. A hole that is out of round; egg-shaped.

End-Play The distance a shaft will move longitudinally.

Energized A term used to signify a component or circuit that is activated electrically.

EPC An acronym for electronic pressure control.

Erratic A term used when the operation of a component or system is not as designed or is intermittent.

Extension Housing A housing that encloses the termination or output of a component. Like a transmission output shaft and its supporting bearings.

Fail Safe A term used for a backup system designed to protect certain systems on the automobile.

Fail-Safe Mode A special operating mode used to help eliminate complete break down of a system and allow the vehicle to reach a safe location. This is also known as limp mode.

Fault Code Digits generated by the diagnostic programmer as an aid in troubleshooting.

Final Drive The differential gears that provide power to the drive wheels.

Flexplate A slightly flexible steel part transferring power from the engine crankshaft to the torque converter.

shifts and down shifts.

Ground A return path for an electrical current.

Ground-Side Switched A method of controlling current in an electrical circuit on the negative side of the component involved. Commonly used with electronic systems.

Half-Shaft Joint A term used for a universal joint in the half shaft.

Hang-Up A term used when something malfunctions, usually temporarily.

Heat Exchanger An apparatus that allows heat to be transferred from one medium to another, using the principle that heat moves to an object with less heat.

Hot-Side Switched A term used for the method of turning electrical and electronic circuits on/off from the positive side of the circuit.

Idle Speed Speed of the engine in rpm, at curb idle, under no load.

Ignition Switch The main electrical switch of the vehicle, includes off, start, and run positions. Usually also includes an accessory (acc) position.

Inputs A term used with various electronic systems to identify the different signals sending information to the electronic modules about operating conditions.

Inch Pound A unit of measure of torque in the English system.

Intake Manifold A metal component used to direct and duct the air fuel mixture into the cylinders.

Intermittent A condition that occurs at random rather than as designed.

Lapped A process of lightly smoothing or dressing a surface to check for straightness, flatness, or accuracy.

Line Pressure The base pressure established in a transmission/transaxle by the pump and pressure regulator valve.

Load The demand for power placed on an engine.

Lockup Converter A hydraulic torque converter that has friction material and a disc that can lock mechanically when commanded by an electronic module to improve fuel economy.

Lockup Solenoid An electrical solenoid that can enable or prevent torque converter lockup.

Neutralizing The centering of shafts and other structures to eliminate stress and/or rotating problems.

Nipple A term often used to signify a place to attach a fluid line, vacuum line, or vent hose.

Normally Open (NO) A term used to signify the usual operating status of electrical devices or fluid control devices. When the proper command is received the device switches to closed.

Ohmmeter An electrical device used to measure resistance.

Oil Cooler A radiator- or tank-like device used to cool engine, transmission, power steering, or other fluids.

Parking Pawl A mechanical device required on automatic transmissions and transaxles to physically lock the output shaft and hold the vehicle in place when the shift lever is placed in park.

Planetary Gear A gear set providing the means for neutral, low, medium, high, and reverse operations.

Powertrain Control Module (PCM) The industry standard term for the computer used to run the engine controls and transmission controls in most applications.

Preload A specified pressure applied to a part or assembly during assembly or repairs so the part can operate as designed.

Pressure Test Varied methods on various systems to determine if components can hold pressure, or if the proper pressures are being developed and controlled in different automotive systems.

Radiator A heat exchanger in the engine cooling system.

Relay An electro-mechanical switch.

RPM/rpm Acronyms for revolutions per minute.

Scanner A common name for various tools that can access information from different electronic control modules controlling electronic systems.

Selective-Thrust Washer A special thrust washer that is furnished in various thicknesses so clearances and preload can be adjusted and changed if necessary.

Short Circuit The intentional or unintentional grounding of a current carrying electrical wire.

Slips A term often used in the transmission area indicating a condition of lost efficiency. Bands slip, clutches slip, and one-way clutches slip.

Solenoid An electro-mechanical device used to impart a push-pull motion.

Speedometer A dash mounted device used to indicate road speed.

Speed Sensor A device on today's vehicles that is an input to many of the different electronic systems. It usually relays component speed to the proper module with a sine wave type signal.

Stall Test A test performed on automatic transmissions to determine engine and transmission condition.

Throttle Valve (TV) A device connected the throttle linkage or the modulator valve and responds to engine load. The higher the demand the higher the pressure rises. It is often adjustable.

Torque The measure of a force producing tension and rotation around an axis.

Torque Converter A unit that transmits power from the engine to the transmission.

Torque-To-Yield To tighten to a specified predetermined yield or stretch point.

Torque Wrench A tool used to measure and tighten a device to a specific torque.

Transmission Fluid A lubricant formulated for use in transmissions.

Transmission Oil A fluid formulated and designated for use in a transmission.

Transmission Oil Pan A removable part of the transmission that contains its oil.

TV An acronym for throttle valve. The throttle valve is connected to the throttle linkage or the modulator valve and responds to engine load. The higher the demand the higher the pressure rises. It is often adjustable.

VSS An acronym for vehicle speed sensor. A device that creates a sine wave signal used by on board computers to determine vehicle speed.

Warning Lamp A light placed in the passenger compartment to warn the driver if a problem in a monitored circuit occurs.

Notes

Notes

Notes

Notes